Landscapes in Music

WHY OF WHERE

Series Editor: George J. Demko

The goal of this series is to provide new perspectives on important ideas and issues of our time in a relevant and lively style. Topics such as peace, migration, health, death, genius, music, homosexuality, and other global themes are explored across space and through time, highlighting relationships, patterns, and concepts. Intended to educate *and* entertain, these volumes will appeal to students and general readers alike as they demonstrate the significance of the geographical perspective in the world today.

The Chinese Diaspora: Space, Place, Mobility, and Identity, edited by Laurence J. C. Ma and Carolyn Cartier

Landscapes in Music: Space, Place, and Time in the World's Great Music, by David B. Knight

Landscapes in Music

Space, Place, and Time in the World's Great Music

David B. Knight

ROWMAN & LITTLEFIELD PUBLISHERS, INC.
Lanham • Boulder • New York • Toronto • Oxford

ROWMAN & LITTLEFIELD PUBLISHERS, INC.

Published in the United States of America
by Rowman & Littlefield Publishers, Inc.
A wholly owned subsidiary of The Rowman & Littlefield Publishing Group, Inc.
4501 Forbes Boulevard, Suite 200, Lanham, Maryland 20706
www.rowmanlittlefield.com

P.O. Box 317, Oxford OX2 9RU, UK

British Library Cataloguing in Publication Information Available

Library of Congress Cataloging-in-Publication Data

Knight, David B.
Landscapes in music : space, place, and time in the world's great music / by David B. Knight.
p. cm. — (Why of where)
Includes bibliographical references and index.
ISBN-13: 978-0-7425-4115-3 (cloth : alk. paper)
ISBN-13: 978-0-7425-4116-0 (pbk. : alk. paper)
ISBN-10: 0-7425-4115-0 (cloth : alk. paper)
ISBN-10: 0-7425-4116-9 (pbk. : alk. paper)
1. Orchestral music—History and criticism. 2. Music and geography. 3. Landscape. 4. Geography—Songs and music—History and criticism. I. Title. II. Series.
ML1200.K63 2006
784.15'6—dc22

2005023037

Printed in the United States of America

∞™ The paper used in this publication meets the minimum requirements of American National Standard for Information Sciences—Permanence of Paper for Printed Library Materials, ANSI/NISO Z39.48-1992.

Contents

Preface

Just as poets and novelists describe, painters paint, and geographers analyze landscapes, so many composers of orchestral music represent landscapes (and waterscapes) that are real, imagined, or mythical and reveal an awareness of various physical and human processes and attachments to place. For example, landscapes of death, survival, and remembrance exist in music, just as they do on the surface of the earth and as part of our geographies of the mind. This book considers orchestral music in which are represented natural forces, humans, birds and animals, some creatures of the sea, and physical and humanized landscapes. Also considered are certain places used for the performance of music.

It is to be hoped that this book will help readers to see in some orchestral works ideas and interpretations that have hitherto escaped them. The music discussed is organized according to a variety of themes and subthemes that emerged from a study of the music. The appendix, in identifying research themes and publications by those who have studied the geography of music, indicates why this book's focus on orchestral music and landscapes is an exploration of "new ground."

What we "see" in a landscape may or may not be seen by others, so we must be aware of contrasting perceptions of the same landscape and of the cultural framework within which people perceive and use the landscape. We can thus ask, would two composers "compose" the same scene in an identical manner? As will be seen, no! As each composer seeks to put into music his or her perception of a particular landscape, his or her musical style is used to express a representation of that landscape. In that sense, each composition is unique. The same holds true for performances.

No performance of a piece of music is ever repeated exactly as it was before. Interpretations are always different. Musicians in the orchestra and the audience will share *an* interpretation, not *the* interpretation. Music is not static. It is changed from performance to performance by altered tempi, emphases, volume, and more by a conductor and players; hence, the audience may be surprised from one performance to another.

I write about orchestral music from my perspective as a cultural geographer. Accordingly, I hope a new appreciation for geographical insights will result from reading this book. Further, I also hope that the book will help you, the reader, to think afresh about links between composers and the "environments" they create in their music.

The journey taken to write this book began many years ago. When I was about fourteen, my mother bought our first thirty-three rpm record on which was Jean Sibelius's *Finlandia*, one of her favorite pieces of music and, only incidentally, also his Symphony no. 5. Little did she know at the time that the second of these two works was to have a profound influence on my development as a musician. I had just completed six months with a music teacher at George Watson's College in Edinburgh, Scotland, who told me of his visit with Sibelius in a cottage deep in Finland's woods. That story impressed me greatly, and then, some months later, our family owned a record of a symphony by Sibelius. I was in heaven! Three years later, I was given a copy of the study score of the Symphony no. 5, which I studied for months thereafter. Since then I have benefited from chats with and the interpretations of numerous conductors and musicians in orchestras from all over the world. I first played in pipe bands in New Zealand and Scotland. I have since performed numerous orchestral scores during more than forty years of playing timpani. Within a year of my first lesson, I had played timpani with a Scottish Youth Orchestra, and just before I left Scotland, I had the thrill of playing timpani with John Fletcher on organ for a BBC broadcast that was heard both nationally (in the United Kingdom) and internationally. Dr. Cedric Thorpe Davie of the University of St. Andrews and Fletcher, until recently professor of music at the University of the West Indies, encouraged me to study orchestral scores. What marvelous advice. I will never forget their interest, generosity, and encouragement. I then moved to the United States where I was hired as the drumming instructor at Macalester College in St. Paul, Minnesota. While in St. Paul-Minneapolis, I played timpani in two chamber and two symphony orchestras. Later, in Chicago, while working as an editorial geographer for the *Encyclopaedia Britannica*, I played with the famous Chicago Business Men's Orchestra, the Northwest Symphony, and various orchestras for ballet and opera, including the Clarion Symphony Orchestra and Verdi Opera Theater in Detroit, Michigan. In Canada—the wanderlust continued—I have played in several orchestras, including the

Guelph Symphony Orchestra, the University of Guelph Orchestra, and Toronto's Hart House Orchestra. I have also performed in symphonic bands, including the Guelph Concert Band, the Toronto Teachers' Wind Ensemble, and in Montpelier, Vermont, the Capital City Band. I have also performed with the Vermont Philharmonic Orchestra.

I am also a geographer, having completed my Ph.D. at the University of Chicago. I have been on the faculties of Macalester College, Eastern Michigan University, De Paul University, Carleton University, and the University of Guelph. In visiting capacities, I have been at universities in the United Kingdom, Israel, Russia, and the United States. While at Carleton University for twenty-two years, I offered a basic cultural geography course, which included lectures on the geography of music. Students from the Department of Music would join the class for those lectures. I also gave an advanced course in which one of the themes considered was the geography of music. I remain grateful to my colleagues for permitting me to give that exploratory course and to the students in both classes for their interest. This book has thus grown out of those teaching encounters, plus some of the many experiences I have had as a musician.

SOME DEFINITIONS

The following definitions may assist the reader who is not conversant with orchestral music. The most complex of musical compositions is the *symphony*, which is a large composition for orchestra that is divided into three or four distinct parts, or movements (such as a fast movement, a slow movement, a minuet and trio—or a lively scherzo—and another fast movement), each with its own theme or themes. A *concerto* is an extended composition for a solo instrument and the orchestra, usually with several movements. A concert *overture* is a short, one-movement work of distinctive character for orchestra that may be used to open a concert. It is derived from the opera practice of having an orchestral prelude to an opera. A *tone*, or *symphonic*, *poem* is a piece of orchestral music in one long movement that often follows in its development a story, with various themes being repeated and interwoven. A *mass*, as used in this book, refers to music written for a church service but often played in concert. It consists of five large sections: "Kyrie," "Gloria," "Credo," "Sanctus," and "Agnus Dei." The term *woodwinds* most commonly refers to the oboes, cor anglais (or English horn), flutes, clarinets, and bassoons. *Brass* is commonly used to refer to the French horns, trumpets, trombones, and tuba, whereas the term *strings* refers to the combined forces of the violins, violas, cellos, and double basses. *Timpani* (the commonly used Italian name) refers to the two or more large hemispherical drums covered with stretched calf hide or a variant of plastic, which can be tuned precisely (and, on modern instruments, quickly, using a pedal mechanism) to different notes. They are also known as *Pauken* (German), *timbales* (French), and sometimes, in Britain, as kettledrums (Montagu 2002). The term *percussion* refers to the orchestra's "kitchen department," which includes such items as snare or side drum, bass drum, cymbals, xylophone, glockenspiel, tubular bells, gongs, castanets, and so forth.

Acknowledgments

Given the book's gestation period, the list of people to acknowledge is long. Some of the people in music who have had a major impact on my thinking, in some instances surely more than they will ever know, include Vera Gilbert, Bruce Mulholland, Cedric Thorpe Davie, John Fletcher, Miss Afleck, Tom Nee, Herman Straka, Leopold Sipes, Stanislaw Skrowaczewski, Charles and Nancy Boody, Mario Bernardi, Gabriel Chmura, R. Murray Schafer, Henry Janzen, Colin Clarke, and the many students at Carleton University who got excited about the geography of music, most notably Stephen Thirwall (1992). Three of my geography professors from many years ago deserve to be acknowledged for so willingly sharing their excitement about their own scholarly research and for encouraging and nourishing my desire to tackle work at overlapping disciplinary "edges": Hildegard Binder Johnson, Alan C. G. Best, and Marvin W. Mikesell. My gratitude to them is immense. For their long-term friendship and interest in my research, I gratefully acknowledge John Clarke, Iain Wallace, Tom Wilkinson, and the late Duncan Anderson. Some of the research done for this book was made possible by research grants from Carleton and Guelph universities.

At my request, Alan Best and Stanley Waterman (both geographers) and Robert Evans (a composer and former music educator) kindly read and critically commented on the manuscript in draft form and made many valuable suggestions. Though I was not able to accept all of their advice, I am deeply indebted to them for having helped me in significant ways. I am further indebted to Evans for generously discussing with me the issue of representation in music and for reacting to some of my interpretations. In addition, Evans has kindly permitted me to use some of his writings associated with his compositions. My now deceased father, Professor George A. F. Knight, read an early draft and provided cogent reactions from New Zealand. My sister, Ann Knight, did likewise, demonstrating a knack for finding errors and giving fruitful feedback. Professor Kok-Chiang Tan generously shared his knowledge of Western orchestral music in China. David Parsons shared insights about some Canadian composers. The resources of the Canadian Music Centre and the Australian Music Centre were invaluable. Pascale Parenteau of the Centre for New Zealand Music provided valuable information and put me in touch with composer Gillian Whitehead, who, in turn, generously responded to my questions about her compositions and other matters. Elizabeth P. Roberts and several colleagues of the Boston Symphony Orchestra and Frank Villella of the Chicago Symphony Orchestra kindly helped me in various ways. Joanne Martin of the Elora Festival kindly gave me data on that festival.

Over the years, librarians in Canada (Carleton University, University of Guelph, Wilfrid Laurier University, and Centre Wellington Library), the United States (University of Vermont and Norwich University), New Zealand (National Library in Wellington, Universities of Victoria and Otago, and the Dunedin Public Library), and the United Kingdom (University of Cambridge) have guided me to source materials. During research periods away from the University of Guelph, I have been given research privileges at Norwich University in Northfield, Vermont. Esther Farrell, a talented musician and editor, helped me make sense by sharpening my writing. Conny Davidsen helped me formulate figure 1.1 and prepare the index. I am grateful to Raymond Braida for believing in me.

At Rowman & Littlefield, I am grateful to Susan McEachern, editorial director, and her assistant, April Leo, for their patience, support, and assistance. I also happily acknowledge the fine skills of the publisher's copyeditor, Jen Kelland.

I am especially indebted to George Demko of Dartmouth College. His excitement about geography and music is infectious, and his encouragement as I wrote this book is deeply appreciated. As ever, if there are mistakes in this book, they are mine and not those of any of the persons just listed.

For her understanding of my need to be active in music as well as be a geographer, I dedicate this book to my wife, Janet, with gratitude.

1

Introduction: Soundscapes, Geography, and Music

SITUATING THE CENTRAL THEME

Landscapes in Music links music and geography. With landscape as the unifying concept, the text explores orchestral music that represents actual and imagined physical and cultural landscapes, natural processes, and people and wildlife. Contrasting forms of orchestral music from several centuries and by numerous composers of different national origins are brought together and discussed according to major themes (identified by the titles of chapters) and subthemes (identified within each chapter). While a lot of music is not readily perceived in geographical terms, it will be demonstrated that with a bit of digging, it is possible to reveal geographical linkages.

This book is not about tastes in music. Nor is it my intention to present a dissertation on the inner workings of music as an academic musician might in preparing a learned treatise for a specialist audience. For that, the reader can turn to biographies of composers and detailed analyses of specific works. Rather, my purpose is to discuss how composers have explored various themes and subthemes and thereby contributed to the general theme of the representation of landscapes in music. This chapter situates the book in the context of music and geography by discussing, first, soundscapes, the concept of landscape, and shared fundamentals. Thereafter, music is discussed from a variety of perspectives, including the sticky issues of representation, meaning, reactions, styles, use of words, and interpretation.

SOUNDSCAPES

Sounds are all around us, if we stop and listen. Sounds of nature and from human activities create interwoven soundscapes. As people listen, they may hear a distinctive rhythm or thrill to a sound that captures the imagination. Thus, for instance, the mix of birds' songs in the late afternoon, perhaps a pack of wolves howling at night, or a neighbor with a lousy voice trying to sing an aria from an opera while in the shower with the window open may be enjoyed for their distinctive sounds. Such can also happen with the pounding of a factory press or shriek of a train whistle. But since people attribute meaning to sounds, supposedly value-free sounds may remind a listener of something else. Sounds thus may come to represent something more than what they actually are. For example, birdcalls and wolf howls likely represent "pure" nature, though in the case of wolves, it may be feared nature. In contrast, the factory press and the train whistle probably symbolically represent industrialization. And since hearing a particular sound can evoke memories of some good—or nasty—experience, added meanings may well up in the minds of listeners whenever the sound is heard. Timing is likely a factor in how we respond to particular sounds; hence, a bird singing loudly outside a bedroom window when we are trying to sleep in will be bothersome, while that same bird's song may be pleasurable when listened to at dusk.

The womb is the first environment within which we hear sounds. Not only is the mother's heartbeat significant, but as the embryo grows in the womb, it also hears external sounds. The jury is still out on how an embryo is influenced by the playing of Wolfgang Amadeus Mozart or some other composer's work. It is known that the unborn hear, and after birth, they are able to respond to particular voices or sounds (such as a viola or a singer's voice) first heard while still in the womb. The point here is simply that we are introduced to "environmental" sounds very early on.

The earliest examples of what later came to be known as music started when early hunters and gatherers, presumably as they heard certain sounds in nature, granted them meaning and then sought to replicate them using song, or perhaps by making sounds. A reed flute may have sounded like a singing bird, and a ram's horn like an animal crying out in triumph or pain. Mimicking sounds heard "in the wild" may have assisted the people while hunting or, more likely, while talking about having hunted. To be regarded as music, the sounds created had to assume a different dimension from those heard in nature. As the centuries passed, there occurred a long series of processes that together formed the so-called Neolithic, or agricultural, revolution when plants and animals were domesticated. There followed the development of crop and food surpluses, urban growth, the creation of mathematics and writing, leisure time, and occupational diversity—including the identification of specialist musicians. It was then that music began to take organized form and started its still ongoing development.

The term *soundscape* was concocted and defined by the composer R. Murray Schafer (1977, p. 274) as, simply, "the sonic environment" and was proposed to encompass all sounds around us. The term can include music written for out-of-doors presentation (see chapter 8) and music performed indoors in which composers, drawing their inspiration from sounds in nature, incorporate those sounds into their compositions (Gilman 1966). Composers may be inspired by a soundscape, and when an orchestra performs their work, new soundscapes are created. The concert hall soundscape is purposeful and yet fleeting. When orchestras create a soundscape, a memory of that musical soundscape or the feelings evoked during an unfolding performance may remain, but the soundscape itself will have disappeared, even as it was created, its "life" lasting but a second or two (depending on the resonance of the hall).

I associate the word "soundscape" with "landscape," in the manner shown in figure 1.1. The word "landscape" has two reference levels, namely, landscapes as sources of inspiration and landscapes as represented in music. Some music describes or uses landscape sounds, be they sounds of human origin or from nature. How these are reproduced or represented, structured, and symbolized varies. Some music refers to landscapes by using human associations, as with music relating heroic actions in a problematic environment; hence, they are referenced landscapes. Still other music has a direct relationship with where and when it is performed; hence, we can refer to the performance landscape. These associations are explored throughout the book.

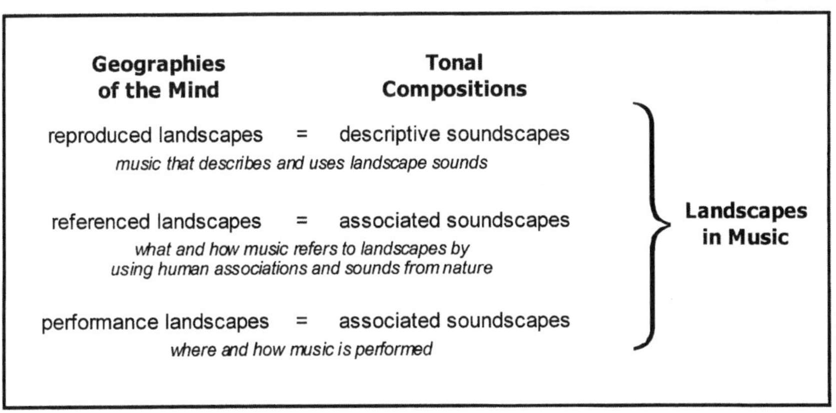

Figure 1.1. The Relationship between Geographies of the Mind and Tonal Compositions, and the Creation of Landscapes in Music.

Ludwig van Beethoven's Symphony no. 6 (known as the *Pastoral Symphony*) is a geographic essay with its descriptions of landscapes, the flowing brook, distinctive birds singing, people at play, a storm like no other in music, and a resolution as peace returns to the countryside. You might dismiss parts of the work as little more than musical geography due to its use of the identifiable sounds of particular birds, including the cuckoo, but unlike, say, the use by some composers of castanets to allude to Spain, Beethoven's birds have a descriptive quality that takes the listener beyond the obvious and into an active world of wonder and change. His pastoral scene and the storm add to the fullness of his landscape in music.

GEOGRAPHY, GEOGRAPHERS, AND MUSIC

I am both a geographer and a musician. This fact alone does not reveal what it is that brings music and geography together. Both geographers and composers describe or represent real, imagined, and mythic landscapes. Geographers use pictures and diagrams, maps, and words to capture the images and symbolism of what is seen and experienced; composers write down notes for instrumentalists to play. Each observes, feels, reflects, and produces "texts." The composer, more than most geographers, likely also reflects on the meaning of his or her work for the inner soul and then lets the reflections influence the resulting composition. The geographer has generally kept such matters out of his or her writings. However, some geographers are now brave enough to "go public" with their very personal insights, questions, and comments on their "truths" (Tuan 1999).

As discussed in the appendix, geographers of music write on numerous themes. Their work is linked by a shared geographical concern for such fundamental concepts as site and situation, spatial distribution, spatial diffusion, and culture regions. My research is framed within an approach pioneered by John K. Wright (1947) and later delineated as humanistic geography (e.g., Tuan 1976). Wright coined the term *geosophy* to refer to the study of geographical knowledge. This book offers a musical geosophy. To take this thought further, I build on Yi-Fu Tuan's observation that "humanistic geographers want to know how a people, not only by imaginative action but also through enlivening speech [and, I would add, musical composition], create and recreate their multifaceted and kaleidoscope worlds" (Tuan 1978, p. 372).

This study explores Tuan's stated concern from the perspective of composers and their music for orchestras. My perspective also draws upon the notion of "geographies of the mind," a phrase coined by geographers David Lowenthal and Martyn Bowden (1976). These and other scholars made geographers aware of the benefits of addressing people's mental images of the world around them, which, in varying ways, reflect attitudes, values, and perceptions that, in turn, reflect an understanding of, and have an influence on, their daily interactions with the environments in which they live (e.g., Downs and Stea 1977; Knight 1983, 1987; Foote, Hugill, and Mathewson 1994; Foote 1997) or, as some will have it, the "ordinary," or "vernacular," landscape (Meinig 1979; J. Jackson 1984). These images also form part of people's identities and senses of group self and are reflected in how "we" relate to "others," on local to national and international scales (Knight 1999a, 1999b).

Concern for attitudes, values, and perceptions underscores the basic notion that nature itself depends, in part, on the contents and assumptions of numerous and diverse geographies of the mind. Physical geographers, biologists, and geologists focus on the biophysical attributes of nature and the many processes that cause change, not on how people experience nature. Cultural geographers, in contrast, identify how people believe nature and landscapes to be and how the latter are represented. Composers clearly are in the tradition of writing geographies of the mind, not providing scientific explanations.

LANDSCAPES IN MUSIC, OR NATURE IN MUSIC?

This book is entitled *Landscapes in Music*. Landscape is important within geography, for not only is it the surface on which we live, but it represents the interface between human thoughts and actions and the biophysical environment. The term *cultural landscape* refers to the landscape that is the result of long- and short-term human (intentional and accidental) decision making, which, in varying ways, changes the physical environment or changes thoughts about the physical environment. The concept of landscape encompasses the term *environment* because a landscape reflects myriad interwoven, natural and human processes that combine in varying ways to create distinctive patterns on the world's surface. I am not concerned with the processes that physical geographers study, although, as will become evident, representations of water and wind and other natural phenomena, and, indeed, more broadly, nature, are to be found in some orchestral music. Rather, this book is primarily concerned with meanings attributed to and reflected in the landscape that have been captured or, at least, are represented in music as expressions of numerous composers' geographies of the mind. Accordingly, the definition of landscape by Denis Cosgrove and Stephen Daniels (1988, p. 1) addresses my focus on landscapes and music: "a cultural image, a pictorial way of representing, structuring or symbolizing surroundings."

Just as geographers and others can have difficulty reading and representing a common landscape with any degree of clarity due to different cultural frameworks and contrasting types and levels of understanding

(Shute and Knight 1995), so, too, do composers perceive, appreciate, and thereafter represent landscapes in contrasting manners. See, for instance, the works on the Antarctic by Ralph Vaughan Williams and Peter Maxwell Davies (chapter 6).

This book could have been titled *Music and Nature*, but since the concept of cultural landscape necessarily includes both humans and the biophysical environment, the concept of landscape is the prime theme within this study. Even so, the term *nature* is used in this book. However, it is not easy to answer the question of when something is "nature" and when it is not. Nature exists only because humans have identified and named "it" as such. Nature and, relatedly, wilderness are as much human creations as they are physical ones, for any meanings they hold for people can and will differ from one culture to another and from one time to another (Evernden 1992; Graber 1976). When examining nature, myriad processes related to three geographical fundamentals are evident, namely, ecological interactions, historical and present-day distributional patterns, and human-environment impacts. Some people (as with traditionally-based indigenous peoples around the world) see humans as part of nature, while other people (especially in Western urban-industrialized states) see humans as standing apart from nature. Such differences can have a profound impact on the geographies of the mind and music. To illustrate, the Chinese have long struggled to regulate nature, encouraged by their belief that the universe is orderly. In the past, more than today, they dreamed of harmony on earth, to balance cosmic order, by controlling nature (Tuan 1998). Chinese composers have had a long-standing tradition of representing landscapes and nature in their music.

There was a long-standing acceptance in the West of nature as a "directive" or "determinant" of human actions (environmental determinism). Then, early in the twentieth century, there was the realization that nature "offers" what we might refer to as "options" that are linked to a society's organizational abilities, infrastructure, technological achievements, and perceptual abilities. Nature necessarily, thus, will have different meanings depending how one "sees" nature in relation to collective assumptions and behaviors. But, perhaps we need not get too tied up in such debates, for our intention is to discuss music written by composers who incorporated *their* understanding or image of landscapes and nature in their compositions for orchestras.

PERTINENT TERMINOLOGY: SPACE, PLACE, AND TIME

Geographers and composers are concerned with space, place, and time. First, *space* is the palette for life on earth. Linda Graber puts it well when she states that "space is not a homogeneous mass, nor the sum of innumerable parts. . . . [S]ome parts of space are wholly different from others" (1976, p. 76). The cultural geographer Eric Isaac suggested that the first instance of differentiation within amorphous space was accomplished by whatever small group of people set aside a small part of their hunting area for use by their god or gods. He referred to that location, that site, as "God's Acre" (Isaac 1964–1965). A site has a specific, or absolute, location (for it cannot be moved, though its internal characteristics can be altered). A site also has a relative location, or situation, which indicates how a site lies in relation to other sites. A site is sometimes referred to as a locale or place.

Place is a part of space that becomes special when people attach meaning to it or obtain meaning from it (Tuan 1977; Sack 1980; Herb and Kaplan 1999). Meaning imputed to place can over time be special, or it can be mundane. We often invest special meaning to place and gain meaning from it because, as Tuan (1977, p. 4) has suggested, "places are centers of felt value." In other words, place is special to "us," whether place is defined from the perspective of an individual or a group (at scales varying from small to huge, such as from a household to a nation). Composers obviously also consider place to be "centers of felt value," judging from the many instances in which place is described in music, as demonstrated in this text.

Time is essential in music, though some contemporary composers, as minimalists, seek to have time essentially stand still. Interestingly, some geographers also made time stand still when they examined a place from the perspective of a time and space "slice." However, the geographers' desire to make time stand still (using cross-sections) proved inadequate for explanatory purposes, so new approaches were tried (Mikesell 1969; Murphy and Johnson 2000). Overwhelmingly, geographers are today concerned with the passage of time via processes (i.e., physical and human processes that cause change in the geographical environment). Composers' concern for change through time is also evident, but sometimes at a scale more intimate than the scales used by most geographers. Notes differ in time, groups of notes are played according to assigned time signatures, music takes time to perform, and so on. (See chapter 2 for some additional comments about time and space.) Composers and geographers also share an interest in how time is understood, though differences in approach make close interdisciplinary analysis difficult.

The best composers write for themselves, in the local setting, which may be totally abstract, purely of the mind, or defined by some suggested source material (such as a scene visited, a painting, literary references, songs, or myths). A composer also draws upon musical and broader cultural traditions and expectations. Just as a geographer might describe a scene prior to analyzing it, so might a composer write about such a scene, perhaps with special consideration of changing colors as the day proceeds. A geographer may be sensitive to the historical context of the study site; a composer may compose a work that includes references to the past to illustrate the present.

The importance of *scale* will vary for the composer as it does for the geographer. To one person, the scale of focus may refer to a favorite pub or maybe a village, whereas to another person, the scale of reference may be a large city or perhaps a broad region within the state. It is important for the composer, as it is for the geographer, to know the spatial limits of the "place" and, thus, to be aware of scale, for the degree of detail—in the geographer's written description or the composer's composition—varies according to the scale of reference and "analysis." As will become evident, composers are well aware of scale. Thus, for example, Japanese composer Tōru Takemitsu (1930–1996) was concerned with specific features and specific places on an intimate scale, whereas Richard Strauss (1864–1949) could be concerned with details but generally was more concerned with the overall large place under consideration on a grand scale. Tuan (1998, p. 175) suggests we need both general space and specific place: "the space of vulnerability and freedom" and "the place of stability and confinement." These contrasting perspectives on space and place are evident in lots of music as a tension.

Ties between the sacred and places are evident in both music and geography. Drawing from Mircea Eliade (1961), sacredness can be attributed to or accepted from territories and places that are set apart from the profane world around them. Such sacredness generally is associated with power. Thus, space associated with sacredness assumes elements of power that are not found elsewhere. Sacredness may be associated with particular phenomena, such as mountains and springs or a national territory, and can be the basis for a search for meaning in life and definitions of national identity. These points are raised here since the ties between sacredness, place, territory, and group identity (especially national identity) have been a source of inspiration for numerous composers. We turn now to the issue of landscape as inspiration.

LANDSCAPE AS INSPIRATION

Geographer Tuan (1998) suggests we need to escape from the mundane. The Finnish composer Jean Sibelius would go to a remote cottage in the woods where he could compose. Gustav Mahler did likewise: he loved the openness of the countryside, as opposed to the sometimes oppressive nature of Vienna society, and he composed in a cabin in the garden of his summer house. Indeed, he had three such cabins over the years.

Confronting Silence is the title of a fascinating collection of essays by Takemitsu (1995), and "Dancing with Silence" is Canadian percussionist and longtime performing collaborator John Wray's description of Takemitsu's compositions. The evocative images these words convey are worth considering before hearing any of Takemitsu's works. A good place to start is his *How Slow the Wind*, a stunning work for its creation of images of gently blowing grass and flowers amid rocks and trees, or his *Rain Coming* and *Water-ways*, both remarkable for his ability to create images of flowing water. Finnish conductor Jukka-Pekka Saraste (Crory 1996, p. 37) said of Takemitsu's work, "All of his music is atmospheric, scenic and related to landscapes and nature." Inspired by nature and finely textured Japanese gardens in which every feature has a special, even precious, and certainly meaningful place, Takemitsu said he wrote "music by placing objects in my musical garden, just the way objects are placed in a Japanese garden" (Crory 1996, p. 37).

What generates inspiration? Why is it that two composers looking at the same garden will then compose quite different works? The mind works in wondrous ways, not all of which are understood. We know, however, that thoughts, feelings, and inspiration can be stimulated by interactions between people and nature, by the influence of culture, by memories, stories from childhood, love for another, and so on. Sometimes a generalized response occurs, as if one's gaze over a broad area leads to a synthesizing idea, while at other times reference may be to a specific feature, such as a brook, a pastoral scene, or perhaps a bird singing. Or, possibly, it is some human event associated with a particular place that inspires creation, notably when loss of life and freedom are at stake. A measure of the character of places can be gained from attending to associated artistic works, such as, for example, Robert Burns's remarkable array of poetry and songs about Scotland's people and animals. Do you remember his description of the mouse as a "wee sleekit, cowrin' timorous beastie" as he revealed his deep love of place and of nature? Or Emily Brontë's compelling portrayal of the bleak moor in *Wuthering Heights*, John Constable's powerful canvas representations of the English countryside, geographer James Wreford Watson's (1979) evocative poetry about landscapes and places in Canada, Ansel Adams's breathtaking photographic images of landscapes in the United States (Gray 1995), or, to bring us back to music, Aaron Copland's delightful representation of spring in the Appalachian Mountains? Consideration of their works reveals how each artist obviously developed deep insights about the places he or she lived in or otherwise saw, the real and imagined people in them, the places themselves as locations (whether specific localities or generalized areas, and whether "good" or "bad"), and nature itself. The process we call inspiration is still poorly understood, for it is not always clear how the stimulus of, for example, seeing a mountain can be translated into a work in music. Yet, we know that involvement with the landscapes around us can have a positive effect.

Many inspirational sources are known. For example, a morning walk in St. Petersburg spurred Pyotr Ilich Tchaikovsky in 1869 to write *Romeo and Juliet* (Hanson and Hanson 1965, p. 99), and a visit to a specific lake in Africa inspired English composer Sir Michael Tippett to compose *The Rose Lake* (see pp. 56–57). Canadian composer, R. Murray Schafer (1984, p. 62) was struck by Canada's "spectacular and terrifying geography" below him while flying: "Immediately I had an idea for a symphonic work in which sustained blocks of sound would be fractured into splinters of color, like sun-glinting off vast folds of arctic snow"; *North/White* was the result. The beauty of surrounding landscapes and sounds within them inspired Austria's Johann Strauss to include folk tunes, hunting horns, and bells in his many works. *The Blue Danube* waltz, his most famous work, was inspired by the glorious river. Russian Anton Rubenstein was inspired by his country's remarkable diversity when, for *Russia*, he drew upon motifs that represented the various nationalities. And when a famous conductor, visiting Mahler in one of the cabins in which he composed, commented on the view, Mahler replied, "I've composed that!" Additional examples are noted in the ongoing text.

It is useful to record that Beethoven said, "I always have a picture in mind, when I am composing, and work up to it" (Thayer 1967, p. 620). His use of the phrase "picture in mind" does not necessarily refer to a picture of a landscape, person, or event, of course, but to a "picture in music" as he wants it to be. Still, this image is a good one, for it speaks to the presence of *inspiration* (that something that triggers thought), *imagination* (the spark for actually composing), *construction skills* (creating in music written on a page an understanding of what is "heard" in the mind), *memory* (as the composer remembers what the total "picture" is to sound like), and *orchestration* (or the "handling of the parts" by different instruments). Sometimes Beethoven did have specific "pictures" in mind: his *Pastoral Symphony* includes a reference to a particular stream in Vienna's woods (see p. 75).

Like any writer, a composer has to be methodical when composing. A composer friend of mine has a clothesline strung across his studio, and on it he places "thoughts" and "fragments" that will be used as he works. This image is wonderful, for it reminds us that composing is an ongoing business. There are few Mozarts. He had the remarkable ability to create a work fully in his head and then write it out cleanly. Most composers have to take time, scratch out, rewrite, even start again, and redraft yet again as thoughts develop—in the manner of many students while writing papers (Northey and Knight 2005). Beethoven's manuscripts show that this was his approach; some of his scribbles as he changed his mind about this or that are hard to decipher. And there is the delightful story of Johannes Brahms being asked one evening how he had spent the day. "I was working on a symphony," he replied. "In the morning, I added an eighth-note. In the afternoon, I took it out."

American composer Ellen Taaffe Zwilich (1939–) can help us regarding the process of composing. She spends a long time thinking before starting a new work. "Inspiration engenders product," she says, "which, in turn, engenders more inspiration . . . but it must be *sound*." Further, "once I'm writing, something mysterious happens. . . . [O]nce you're into a new work [you] let it take you somewhere you've never been before" (*New York Times*, July 14, 1985, quoted in Jezic 1988, p. 177). Some composers, including Brahms and Anton Bruckner, knew they were inspired by God to write all they composed, not just their explicitly "religious" music (Abell 1994). And Sibelius talked of "wrestling with God" when he revised his work. From somewhere—internally or externally—inspiration occurs, but perhaps the true source will always remain a mystery. The key question may well be, is the person open to accepting inspiration, whatever its source?

Before proceeding, it is useful to focus on music by providing a definition and discussion, briefly reviewing Western music, and considering issues pertaining to meaning and representation, using words to describe music, composers' autographs, and the "Landscape of the Score" so that the reader can better situate and appreciate the many works discussed in this book.

WHAT IS MUSIC?

Music is the art of organizing tones to produce a structured, coherent, and repeatable sequence of sounds that form melodies and harmonies played in varying rhythmic patterns by performers on instruments of varying types, all of which can elicit intellectual and aesthetic responses in a listener. This rather formal, even intimidating, definition deserves to be simplified. Helen Kaufmann defines music as "the art and the science of expression in sound." An equally pithy, less formal, but still provocative comment by an anonymous writer is especially useful: "Music is the art of sounds in the movement of time" (quoted in D. Watson 1994, p. 3). These two added words—movement and time—have special significance for the theme of this book.

Music is intended to be played, listened to, experienced, pondered over, reflected on, and, ultimately, enjoyed. Music speaks to several of the senses, not just hearing. We cannot literally taste music, but we have tastes in music. We cannot catch music, but music can catch our attention. We can feel music—and not just from throbbing speakers in teenagers' cars as they drive by with the radio blaring—for all music, when performed, sends vibrations through our bodies, and we often respond to music by saying what we feel about it. Music can challenge, inform, remind, excite, agitate, sooth. Music is for all ages, and all ages have provided music, some of it for "all time." Some music cannot literally be repeated, being based on never-ending improvisations; other music can be repeated, as performers play from memory or from what has been written down for them to read—yet each performance will always remain unique (Barenboim and Said 2004, p. 29). Composers write music that can create all sorts of moods and reflect all sorts of influences, some of which are geographical.

We may *like* music—and here no judgment is made concerning type, form, tradition, or whatever else may influence personal judgment—perhaps because, in British poet W. H. Auden's words, "music is the best means we have of digesting time." Why is this so? Perhaps because, in Boston conductor John Oliver's words, "music lifts us all out of our little lives" and, suggestively, in Charles Baudelaire's words, "music fathoms the sky" (Auden and Baudelaire are quoted in D. Watson 1994, p. 3; Oliver is quoted in Vigeland 1991, p. 47). These last comments are not definitions, but, as statements, they are nicely poetic, and perhaps that is how most people respond to music.

Simply *enjoying* music, because "I know what I like," is one way to approach music. Other perspectives include a concern for certain characteristics. For instance, David Willoughby identifies the concern for music's acoustical characteristics. This involves the study of music in the environment within which it is performed, since music, as the science of sound, has a physical basis. Music is also an aural phenomenon; people will respond to its sounds and silences. There can also be a concern for music as movement through time since it involves a forward movement with, as Willoughby (1996, pp. 13–14) states, "energy, a momentum, a predictable progression to a clear conclusion, such as the end of a phrase" or the end of a movement (i.e., a section of a work).

Another set of perspectives identified by Willoughby starts with the assumption that music has cultural, artistic, and functional qualities. Accordingly, music is an art, and because great art can transcend the local, it can have universal appeal. Music is also a means of expression that "can communicate feelings and images and can generate aesthetic responses, responses that may be universal and transcend cultural boundaries or it may be culture-specific (and therefore it remains local, or non-universal)" (Willoughby 1996, p. 14). Further, music is a psychological phenomenon for it "has the ability to affect and change people's feelings and attitudes," and it "changes as the society it serves changes and because people's needs and tastes change" (Willoughby 1996, p. 15). The latter implies that music follows societal change, which is true to a degree, but certain composers have dared to be different by challenging the "expected" and have struck out in new directions, so they have made the first move, not society at large.

Whatever the reason for music, an astonishingly diverse array of musical traditions—involving distinctive melodies, harmonies, rhythms, instrumentation, nomenclature, and meanings—has been developed by different peoples living in varying locations on the vast surface of the earth and who, as such, are of interest to geographers (see the appendix). Musical traditions reflect peoples' various and contrasting cultural traditions and societal expectations and constraints. Influences from one region to the next, due to contact or lack thereof, have affected the way music has developed in different world regions. The various regions were more clearly definable a century ago than they are today because increased global communications and interactions have led to a sharing of musical styles; thus, fewer well-defined boundaries now exist between musical styles, rhythms, and instrumentation. Yet, regional musical styles are still to be found in many parts of the world, the cultural cores and technical bases of which have existed for many decades, if not centuries. All over the globe, people create and recreate sounds with the human voice, diverse instruments, and, in some cases, computers. Some music, including some Western "classical" music, is dismissed as "noise," but that, of course, is a matter of personal preference and judgment.

DESCRIPTIVE, PROGRAMMATIC, ABSOLUTE MUSIC

This book is concerned with how nonmusical features, events, and processes are referred to or represented in music that is either descriptive (clearly representing an image of something, such as a bird call, a mule ambling along, a storm, or a specific event) or programmatic (representing a distinct mood or emotion or depicting a story). Leonard Bernstein (1959, p. 16) identifies three kinds of such music: narrative-literary (e.g., *Till Eulenspiegel*, *The Sorcerer's Apprentice*), atmospheric-pictorial (e.g., *La mer*, *Pictures at an Exhibition*), and affective-reactive (involving triumph, pain, wistfulness, regret, melancholy, and so forth, all typical of nineteenth-century Romanticism). I have chosen not to use his categories for he uses them disparagingly as they pertain to meanings in music. Bernstein believed one further category was the most reasonable and logical way to categorize music, namely, music that has only purely musical meanings.

Music that is entirely free of extramusical association is called "absolute music." "Absolute music is a result which neither composers nor critics can understand until they have cleared their minds of all confusion with the process that lead to it" (Tovey 1941, vol. 1, p. 66). However, as the text will reveal, composers have knowingly written numerous works to be descriptive or programmatic, which, thus, are not absolute music. This noted, American composer Aaron Copland felt that only "simple-minded" souls want to know the "meaning in music" and that they will never be satisfied with the realization that it is simply not possible to "state in so many words what the meaning is" (Copland [1939] 1988, p. 12). The more the music reminds such people of "a train, a storm, a funeral, or any other familiar conception," he continued, "the more expressive it appears to be to them" (Copland [1939] 1988, p. 13). He felt that if any realization of "the world" could be expressed in music, it could only be in a most general way or, as he put it, "no closer than a general concept" (Copland [1939] 1988, p. 13). Despite Copland's comments, crusty stance, and sometimes nasty bite, he too wrote both programmatic and descriptive music, including *Rodeo* and *Billy the Kid*, the latter including a dramatic gunfight.

Richard Strauss felt that there "is good music and bad music. If it is good, it means something—and then it is Programme Music" (quoted in Amis and Rose 1992, p. 15). As will become clear, many well-known and some not-so-well-known composers have intentionally set out to create in their music a musical idea that reflects what they have seen, heard, or read about in and about particular places. As a result, a particular landscape image is evoked, animal and people sounds are suggested, or the telling of a story or relation of an event is revealed by the use of a particular instrument or combination of instruments, a certain tone color, a particular rhythm, or, perhaps, a reference to a folk tune; thus, a listener to the music can hear orchestral sounds that reflect sounds either "in nature" or refer to a particular cultural-historical and geographical setting. In seeking to represent nature and landscapes in their music, composers were constrained differently during the Baroque and Classical periods than later during the Romantic period. The point here is that some composers have chosen to represent in their music specific, generalized, or mythical scenes or events. In short, one has to take Copland's comments with a large pinch of salt.

Many composers name their works or give titles to the several movements; hence, the tie to landscape is made evident by the composers' own words. On occasion, others' names for a work or its sections are identified. Some have become fixed. Some composers purposely do not publish their own titles for either a whole work or its movements. For example, Alan Bush, a British composer, chose not to publish his titles for the three movements of Symphony no. 3 because, he said, they were descriptive of specific places in Britain and individuals in them, so he feared he would be sued for libel. Other composers to have deleted their working titles have included Robert Schumann and, to avoid political criticism and perhaps prosecution, Dmitry Shostakovich (chapter 7).

WESTERN MUSIC

Some cultural studies scholars want us to stop calling Western music "Western" because of the colonial implications of the word. They would have us use "Euro-American" since, they argue, this phrase embraces what they call "a geographical region" comprising Europe and America. Their term is both inaccurate and geographically nonsensical due to what it omits. What and where is "Euro"? What parts of Europe are "in" or "out?" What of Russia? The issue also raises the question, what is "American"? This word varyingly refers to all or parts of North, South, and Central America and sometimes just to the United States. Canada is in North America, but it is not part of America when the term refers to the United States. So what is Euro-America? The term omits numerous musicians and composers who nevertheless are deeply immersed in "Western" music, including many persons (including some indigenous persons) and orchestras in, among other places, New Zealand, Australia, Brazil, South Africa, Japan, and China. In short, it is not helpful to identify a highly questionable or separable "Euro-American" world region.

In good faith (and with tongue firmly in cheek), in an effort to be inclusive, we may wish to substitute the following for "Western" music: music-from-Europe-with-important-inputs-from-Russia-the-Americas-Australasia-plus-of-course-increasingly-East-Asia-and-notably-Japan-and-China-with-important-contributions-from-indigenous-cultures-around-the-globe. Silly and clumsy as this phrase is, it is clearly more inclusive than the term Euro-American. The latter term is not worthy of serious consideration. Accordingly, I continue to refer to Western orchestral music since it is conceptually open to the dominance of the rich Western heritage and development of music and orchestras, using largely Western-derived instruments, and yet inclusive enough to embrace non-Western influences and orchestral playing in both Western and non-Western settings, including China and Japan, as the ongoing text will show.

It may be tempting to substitute the word "orchestral" for "Western," but this, too, would be unsatisfactory since Western music includes many types of nonorchestral music (e.g., trios and quartets). In addition, orchestras in many parts of the world use instruments not found in Western orchestras. Even so, the term *orchestra* is used in this book. Unless noted otherwise, the word as used in this book refers to two types of orchestra that evolved in Europe, namely, the chamber and the symphony orchestras (Peyser 1986; Knight, in press).

MUSIC IN THE WESTERN TRADITION

What we now call classical music in the Western tradition had its origins in religious settings in ancient times, in Greek, Roman (notably Christian), and other domains, and then, as part of "civilization," it spread to and developed further in Western Europe. Eastern (Byzantine) and especially Western (Roman) liturgies made music the servant of religion. For early Christians, "approved" music enhanced understanding of Christian teachings. However, since "the Word" was critical, instrumental music was in time excluded from, or limited in, some worship services, despite the fact that the Bible (i.e., the Old Testament of Hebrew scriptures) is filled with references to instruments and music. St. Augustine admitted to loving music and in AD 387 began his six-volume treatise *On Music*; yet, he also worried that he was emotionally moved by some music, even that sung in church. Some early church missionaries saw that they could mingle "the precepts of religion with sweetness of melody . . . so that those who are still children [will build] up their souls even while they think they are only singing the music" (John Chrysostom, quoted in Grout and Palisca 1996, p. 26; this approach is even today being applied by some religiously supported soup kitchens).

The geographer Clarence Glacken's remarkable review of centuries of changing relationships between humans and nature includes this statement about Pythagorean belief: that "harmonic ratios produc[ed] the *music of the spheres*" mirrored central beliefs in geography and geometry as they pertained to cosmic order, unity, harmony and equilibrium (Glacken 1967, p. 17, emphasis added; for a musician's parallel summary perspective, see Maconie 1993, pp. 84–92). Anicius Manlius Severinus Boethius (ca. AD 480–524) wrote *De institutione musica* (*The Fundamentals of Music*), compiled largely from Greek sources concerned with arithmetic, geometry, and astronomy. Boethius identified three kinds of music: *musica mundana* (cosmic music), which pertains to the "orderly numerical relations seen in the movement of the planets, the changing of the seasons, and the elements"; *musica humana*, which "controls the union of the body and soul and their parts"; and, finally, *musica instrumentalis*, or "audible music produced by instruments, including the human voice, which exemplifies the same principles of order, especially in the numerical ratios of musical intervals" (quoted in Grout and Palisca 1996, p. 29). Music, in other words, had to fulfill certain assumptions about humans in relation to God and the cosmos. This belief in cosmic (and, thus, "heavenly") music held until the seventeenth century, when, in Yi-Fu Tuan's (1998, p. 176) words, "the spheres finally stopped playing."

It is important to consider music as an object of knowledge, not necessarily just of performance. Some people maintain that the "true" musician is not a singer or a composer who uses instinct "without understanding the nature of the medium, but the philosopher, the critic, 'who [in Boethius' words] exhibits the faculty of forming judgments according to speculation or reason relative and appropriate to music'" (quoted in Grout and Palisca 1996, p. 29).

The acceptance of music as something of consequence was useful to kings, princes, and bishops. By encouraging or, in some instances, simply condoning the development of music through official support of composers and musicians, they were seen to be learned men. Also, not so incidentally, they varyingly controlled the development of music because they essentially controlled most composers.

Identifying the twists and turns of music in and outside the courts and the church is not the intention here, but we can note that most serious *musica instrumentalis* were originally performed for the courtly classes in churches and in the courts of power. The "common" people had their vernacular music, music generally denigrated by those in authority. However, the lower classes learned they too could enjoy some of what we now call "classical" music when the music of the aristocracy was made available to them. George Frideric Handel (1685–1759) was undoubtedly the most effective and best-known composer to take his music to the people: his *Messiah* was performed outdoors for crowds of thousands in Ireland, England, and Germany. We may think that music "crossover" is a new concept (of the late twentieth century). It is not. By their actions, Handel and other composers crossed the divide that had separated classical music for the upper classes from the masses with their believed preference for their own vernacular music.

Most serious music had long remained in the confines of the small room for an aristocratic audience, but pressure to provide for the wider audience increased, and the examples of Handel and Mozart represented the start of a revolution of a kind, inasmuch as there proved to be a need for large concert halls (Broyles 1986). Beethoven's music benefited most from being performed in such halls, though he, through his development of the symphonic form, really was the revolution. Indeed, due to him, an increase in reverberations in performance halls was deemed necessary. Music by Handel or Franz Joseph Haydn works well in spaces with little reverberation, but large compositions such as Beethoven's later symphonies, Claude Debussy's *La mer*, or *The Sorcerer's Apprentice* by Paul Dukas are best heard in large halls with longer reverberation times.

The best orchestras in the world perform in the best spaces. Some of the halls are quite old (chapter 8). Musicians in orchestral music range from fully professional to semiprofessional and from excellent to not-so-good amateur. The nature of today's audience is far wider in socioeconomic terms than was the case in the early 1700s, and there are many more attendees at classical concerts today than ever before in history. There may still be a beneficial carryover in North America from the wonderful series of midcentury concerts broadcast live on radio by several orchestras (see p. 204). Many factors account for the strong interest in classical music today. These include the thousands of regular live performances by symphony orchestras, chamber orchestras, and smaller groups of all sizes and with varying degrees of skill.

Some people unthinkingly dismiss Western classical orchestral music because of its sophistication and the complex structures that undergird its composition and performance, even as, ironically, they may learn to appreciate the complexities of Latin American, African, or South Asian folk music. Some Western classical music is easier to listen to than some other Western classical music, and some Western vernacular music is easier to listen to than some Western classical music. However, there is a lot of Western classical music that offers greater insight into the human condition than much of Western vernacular music. What is the issue? In part, there has always been greater interest in vernacular music than classical music. Vernacular music is played "everywhere," whereas classical music is not. With classical music, one needs to take a chance, to listen to music that at first may seem strange, but that may soon grow on the listener. Classical music is at once simple (many works can be heard for enjoyment), yet complex (due to the many rules and necessities that lie behind the composition and playing of a score). It helps to obtain at least an elementary understanding of classical music. When this occurs, wonderful horizons can be opened for exploration. This is partly a philosophical issue, one that pertains to meanings and representation in music.

UNDERSTANDING AND RESPONDING TO MUSIC

How can we "make sense" of orchestral music? This question has numerous answers, one of which might derive from a music-theory course, another from learning how to play an instrument, and a third from attending orchestral concerts. Our special concern here is with meaning and representation in orchestral music and with emotion, intellect, and feeling.

Ballets, operas, and movies are given only passing mention in this book, partly because they each deserve separate consideration, involving as they do dance or vocal sounds (spoken or sung), but also because they use visual "aids" (scenery, acting, costumes, varied lighting) and a story line that is revealed as a work is performed. Music alone is generally not central to the story line in most operas, ballets, musical theater, and movies, though, obviously, it is expected that the music will necessarily be interwoven with meaning and purpose as it explores and enhances the sung, acted, or danced story. In many movies, the importance of music in enhancing the power or subtlety of a scene or otherwise guiding a response may be critical; yet, the audience may not always consciously appreciate it. Indeed, consider how many times you have seen a movie and then been surprised to see the number of items on the list of music credits at the end. Some music in movies is like a bass drum in a good Scottish pipe band in that you are not aware of it until it stops!

Orchestral music must stand by itself. Visual aids are generally not part of the performance (but see p. 205). Orchestral music is to be heard and, if possible, perhaps in unique ways for each listener, appreciated and made sense of. Put another way, using a variant of a now hackneyed phrase, "what you hear is what you get." But is that all there is to it? Is music unidimensional, in which case it might be thought of as purely intellectual? Or is music multidimensional? If the latter is the case, does this imply that an emotional response is possible? Are the intellectual and the emotional necessarily separate? What is it about music that speaks to the heart and soul as opposed to feeding the mind?

The phrase "to make sense" of music may imply that study is necessary before "sense" can be made of what is heard. There is validity to this observation, but a lack of knowledge of musical composition, theory, and performance skills need not hinder the development of appreciation for what one hears. However, *you* may maintain that the music by Tchaikovsky and Antonín Dvořák is "best" though *I* "know" that music by Beethoven and Bruckner is "best"! Such differences of opinion exist because what music one likes is largely a personal matter, though one's cultural and national origins may also have an impact on preferences.

When challenged by a work and trying to interpret it, one must be careful not to read too much into it. If it is accepted that we must learn what the composer is "saying" and not simply rely on what we (i.e., you and I) think or feel is being conveyed, the challenge is thereby presented to us to become acquainted with at least some of the inner workings of music and about how good conductors and musicians seek to let a composer speak through them. By finding out about a work and circumstances around its composition, we can enhance our understanding while a performance is ongoing. But how do we respond to music performed? How do we determine meaning from hearing a work? The answers to these questions are not straightforward. To seek an appreciation for why this is so, we turn to debates in psychology and philosophy.

Some scholars of psychology maintain there is an intrinsic, absolute meaning in music, whereas the others hold that an extrinsic, referential meaning pertains (L. Meyer 1956, 1989). But more, a formalist listener (who values the abstract nature of music for its own sake) and an expressionist listener (who sees music as a means to express emotions) will formulate different reactions. How can we proceed? One way is to ask, should we respond to music with emotion or intellect? There are several philosophical stances, including, for our brief discussion, expression theory, arousalist theory, cognitivist theory, and mirrorist theory (Kivy 1990; Davies 1994). *Expression theory* posits that expressiveness belongs to a work and not to a character in a work and that this expressiveness is thus an expression of the composer, or, more clearly, it is the performers' expression of their concurrent interpretations, emotions, and feelings in the performance of the work (Davies 1994, pp. 168ff.). In contrast, *arousalist theory* holds that emotions expressed in a musical work are actually evoked within the listener and do not exist within the work itself (Davies 1994, pp. 169, 184ff.). Both of these theories are problematic because they imply a one-way relationship. The nonemotion, or *cognitivist*, theory posits that our aesthetic response to music occurs purely on an intellectual plane: if we permit personal associations triggered by the music to interfere with how we hear music, we will obstruct our access to what the composer wants us to hear. Eduard Hanslick (the infamous Viennese music critic who championed Brahms and savaged others, including Richard Wagner and Bruckner) was an early strict cognitivist who felt strongly that attention had to be focused on intellectually appreciated musical beauty, not emotional reactions. Indeed, he believed that instrumental music could not express definite emotions (Davies 1994, pp. 203–20, 281–87). To him, "music means itself" (Hanslick 1963). But not all cognitivists are like Hanslick. Peter Kivy (1990), for example, finds it hard to draw such a strict division between the affective and the intellectual. He accepts the existence of musically based emotions, that is, emotions expressed within the music that can be appreciated intellectually. However, he and others of his ilk hold that while listeners can respond to different sorts of emotions in music—such as joy and sadness—their response is to the sheer beauty of the music itself—and perhaps to the skill of those who performed it—and not to some emotion intended wittingly or unwittingly by the composer. Whatever emotion there is, it is portrayed in the music and, thus, is not an emotion within the listener. Another theory, among several more, Davies states, is *mirrorist* inasmuch as it posits that music's expressiveness can lead to a response of a particular type by the listener. This, too, is problematic inasmuch as people can respond in different ways to some music, just as we have different responses to a book, a painting, or a landscape.

This brief overview of selected theories on emotion and music is useful for its insights into how we, as individual listeners, may listen to, hear, feel, and respond to what is being performed. Which theory is "correct"? Rather than giving primacy to any one of these theories, I believe the *feeling* we have upon hearing music is elemental. Something transpires for most people as they listen to music. It is also often—but not necessarily—the case that a work may evoke an *emotion* within us. I am not concerned with whether the emotion "embedded" in the music and experienced by us, the listener, is derived directly from the music or is triggered by our individual responses to "mere abstractions." I tend toward siding with the latter, however, because any emotion we experience at one performance is likely not the same as that experienced at other times, for our emotional state can vary from day to day. Also, to use extremes, someone in a chronic state of depression will likely hear some music quite differently from another person who is experiencing a manic phase.

Surely, both emotion and intellect are needed if we are to understand and respond to meanings in music. The expressive power of music can evoke an emotional response in the listener, but the form and depth of the response will surely reflect our mood of the moment (stressed, grieving, joyous, and so on), which, in turn, can have an effect on our receptivity. Sometimes we respond purely on an intellectual level to music, perhaps because the composition does not "touch" us, even though we may recognize it as worthwhile or even stimulating, while at other times, emotion, whatever its source, can be overpowering. Further, the setting of a performance may itself evoke a response that, in turn, influences the way we hear and respond to the music. These several theories are useful for analytic purposes but not for the experience of listening to music. Few people go to a concert saying, "I will assume an arousalist theoretical stance tonight, m'dear." Rather, they go because a favorite score is to be played, they are curious about what is on the program, perhaps a particular soloist is scheduled to perform, a famous visiting conductor is to be present, or, simply, because the season's tickets were bought and, well, it's the night to go.

THE DIFFICULTY OF USING WORDS TO DESCRIBE MUSIC

There are skeptics who would surely have thought it impossible to write this book. As the author, I have had to consider how best to describe music using words. Is it possible to use words to describe a medium that relies on sounds? There is a danger in taking liberties, for sure. For example, there is the telling sarcastic comment of Arturo Toscanini, a famous conductor of last century, who, with reference to Beethoven's Symphony no. 3, the *Eroica Symphony*, said, "Some say this [symphony] is [about] Napoleon, some Hitler, some Mussolini. For me it is simply *allegro con brio*" [the marking in the score that means "fast, with spirit"] (quoted in Schonberg 1981, p. 254). To Toscanini, the music alone said it all.

How can one use words to describe orchestral music? This is a tough question to wrestle with, undoubtedly because of philosophers' concern about whether or not it is possible to represent *any* form of art in another medium. Is it possible to read a painting and recreate it in music? Or to do the same for a piece of literature? How can someone adequately represent in music a story, painting, mountain, battle, thunderstorm, or any other source of inspiration? And how best to put into words what is represented in music? You will appreciate the dilemma if you have read a novel and then seen a movie based on the story. You know what liberties the film's writers and directors have taken, and you may not be pleased with the adaptation. The same can pertain to performance in music. For example, six people may all hear a work but disagree on its intent. Or, they may all agree about the message of other music, but could they agree on the words to best describe the work?

Another issue pertains to how to read music for its descriptive content. At once, I must warn you that it is not wise to read too much or indeed anything of a descriptive nature into all works of music. Indeed, to attempt to identify a descriptive musical idea in most works would surely lead to spurious conclusions. Even so, as will be seen, a considerable musical literature nicely incorporates aspects of cultural landscapes and the people living in them, as well as makes fascinating references to certain natural processes and to wildlife.

But again, how should these works be described in words? This is not an idle question. While writing this book, I constantly faced the problem of how best to describe in words composers' compositions. I had to ask myself, have I chosen the most appropriate words to describe that or this work? Am I being fair? Is there a bias inherent in the use of this word or that? If you know the works I discuss, you may ask, how would I have worded that description? Are the words he uses the same as I would use? Or, you may prefer to ask, are his words in accordance with those used by the music critic of a major newspaper? How then did I choose the words used in this book? Some words used are those of the composers, sometimes paraphrased. Other words used reflect my reading of a wide variety of sources, some of which, when located within quotation marks, have the sources identified. Most of the word descriptions are mine. You may find, as I have, that words come quickly to mind as the work of this or that composer is performed; for others, the seemingly easy task of describing their works may be elusive. These differences may be a reflection of how composers write their works, or "autograph," their works. What is meant by this?

COMPOSERS' "AUTOGRAPHS"

Each composer has a distinct and discernable autograph in his or her music. Takemitsu, when asked to compare his music to Beethoven's, said,

"Beethoven's music is like a big, round tree in the middle of a field. My music is like the grass, the flowers and the birds around it" (Crory 1996, p. 37). Beethoven's music has been described as vertical due to the powerful sense of growth toward climaxes; Takemitsu's music, in contrast, can be referred to as horizontal, without necessary growth. Each of these composers' compositions is connected with nature but in quite different ways. Drawing from Neil Crory, if works by Beethoven and Takemitsu were juxtaposed, one would have no trouble identifying authorship: Beethoven created immense sound canvases; Takemitsu created intimate gardens of sound. Each wrote with a distinctiveness of style, and each of their compositions bears its composer's own autograph.

Of the thousands of composers now dead or the many composers alive today, relatively few have been able to achieve the distinction of having their music readily identifiable by numerous people who hear their work. Still, numerous works are recognizable by their distinctive autographs. For example, radio listeners can readily differentiate between *classical* composer Wolfgang Amadeus Mozart (1756–1791), *romantic* composer Richard Wagner (1813–1883), and *modern* (*minimalist*) Phillip Glass (1937–). Or, if you heard brief excerpts from Tchaikovsky's *Swan Lake*, Debussy's *La mer*, and Beethoven's Symphony no. 5, you likely would know instantly which of the three people composed which work, even if the titles of the works did not come to mind.

Composers autograph their music by what and how they write. Using particular combinations of instrumentation, tones, colors, rhythms, and tempos, they create sounds that "belong" to them. Strings and woodwinds may be used a lot, or horns may dominate. Some composers, including Austrian composer Anton Bruckner and Russian Dmitry Shostakovich, are readily identifiable. Shostakovich often used clarinets, trumpets, and snare drum in distinctive rhythms, whereas Bruckner (a superb improviser on the organ) built his symphonies as if they were huge blocks of sound akin to sounds obtainable on the organ, the blocks being linked by delicate passages of intricate beauty. Use of certain tone colors and harmonization set composers apart from each other. For example, there is English composer Frederick Delius's use of subtle woodwind scoring, gently lilting rhythms, and shimmering string playing to provide a distinctive sense of tranquility and spaciousness. Differently, Czech composer Antonín Dvořák's sounds and rhythms were generally his inventions, which are akin to folk melodies imbued with sadness and wistfulness one minute and whirling activity the next. Some composers arrive at their particular styles and establish their autographs quickly; others take time to develop a distinctive style or voice.

Have you ever caught yourself wondering if what you are listening to is by this person or that, until you either stumble across the right composer in your mind or read the CD label or hear the name spoken by a radio announcer? The rapid spread and acceptance of long-playing records, the ability to stack multiple records, then the advent of stereo radio in the 1950s and 1960s, tapes in the 1960s and 1970s, compacts disks in the 1980s and 1990s, and DVDs at the turn of the century, as well as of a new phenomenon, digital radio and recordings—all have led to huge numbers of people being exposed today to an astonishing array of music in ways that could only have been a dream fifty years ago. As a result, more people than ever before in history now know numerous composers' autographs. A case in point is Symphony no. 1 by Zwilich, for which she won a Pulitzer Prize. The author of a successful text on music included her work on an accompanying CD, thereby revealing her "voice" to thousands of student listeners (Hickock 1993, pp. 502–8). Those who do not have access to that text and CD may never get to know Zwilich's distinctive style. Even when fame prevails in one place, ignorance may be evident elsewhere. Such ignorance may be due to lack of exposure in concert programs. Thus, to the surprise of British musicians, few North Americans know of the remarkable symphonies of English composer Robert Simpson (1921–1997).

An issue for many people is their ambiguous feelings toward certain composers. To like or not to like, that is the question! The work of some composers seems to be intentionally ambivalent, perhaps to unsettle the audience. Such can be the case with works by the American John Cage (1912–1992), German Karlheinz Stockhausen (1928–), and Englishman Harrison Birtwhistle (1934–). A hearing of Austrian American Arnold Schoenberg's 1909 *Five Orchestral Pieces*, the United States's Roy Harris's 1942 Symphony no. 3, or new works that premiered in the late 1900s or 2000 and after may upset some listeners today as they did many audience members at premier performances in the past (see Slonimsky 1965, for a glorious collection of invective comments about music, composers, and conductors).

If a work is highly unsettling, we can ask several questions. What is it about me that I am disturbed by the music? Is it the composer's expressed unsettledness or is it my state of mind? This is one of the delightful ambiguities about music: we may never reach an answer, but we can have fun debating interpretations with friends. Some composers are unambiguous in their message: hence, their music has a directness and, sometimes, a symmetrical beauty. Even so, a mysterious element that we cannot quite put our finger on may underlay their directness. For example, what inspired Beethoven to write his remarkable Ninth Symphony? And is theologian Hans Küng (1992, p. 18) correct in thinking that "a mysterious order" undergirds Mozart's compositions? I leave these questions hanging for these are topics explored by others (e.g., Abell 1994; Barth 1986; Hepokoski 1993; Kennedy 2000; Sisman 2000).

INTERPRETING THE LANDSCAPE OF THE SCORE

This section's title is taken from a phrase by composer Takemitsu, who borrowed it from a book subtitle by conductor Hiroyuki Iwaki. Takemitsu (1995, p. 46) states that the conductor's "role is that of a medium standing at the borderline between composer and performer." With this in mind, the phrase "the landscape of the score" is highly pertinent. Conductors read, interpret, and direct (i.e., conduct) music as their prime task. The conductor seeks to remain true to the fidelity of the score. Also, in the estimation of Herbert von Karajan, a conductor "must not only know what comes next but also have a picture of the whole in his mind" (Hallmark 1986, p. 567). To create this "picture of the whole," or the "landscape of the score," the conductor needs to draw upon his or her research, make decisions concerning tempos, and adjust musicians' playing (e.g., ask for louder clarinet or emphasize the strings at the expense of the woodwinds). Each orchestral musician will have an interpretation in mind for his or her individual part, notably the orchestra's soloists (i.e., the first chair players, or principals), but it is the conductor who, in understanding the whole "landscape," opens the musicians to a wider interpretation. A good conductor tries to "never get in the way" of the score. "The music is always there; it is released" (anonymous conductor, interviewed by Partington 1995, p. 91). Interpretation is not simply an intellectual exercise for, as Georg Solti (1997, p. 114) reveals, "you are conducting with your eyes and with your soul." Just as listeners may let music into their souls, so do conductors. There is a difference of opinion about this, however. Composer and conductor Richard Strauss once told Solti "not to get too involved with the music, but to stay somewhere outside it—not to lack passion, but to be dispassionate in the execution." To this, Solti notes that while "this is good advice . . . it's not always easy for me to remain 'outside' the music" (Solti 1997, pp. 79–80). Solti was famous for being "in" the music, as well as for his precision and integrity to the score and for dynamically revealing to the musicians his intellectual interpretation of, and emotional response to, both the music and their playing. As a result, orchestral players knew clearly what was expected of them, and they responded positively. Some conductors are economical in movement, while others are more outward, some to the point of being flamboyant, with the worst drawing attention to themselves rather than to the music. There is the old story about a person who attended her first orchestral performance and was introduced to the conductor afterwards. The conductor asked if she had enjoyed the concert, to which she replied, "I enjoyed your dance!" The gestures have to mean something. Otherwise, communication will not occur. Good conductors communicate effectively with their hands and body language. A conductor's gestures to players form part of the performance landscape for audience members.

The process whereby sounds and images in a composer's mind can be placed as symbols on paper and then read, interpreted, and played on instruments to become music to a listener's ears is miraculous. Some people have the ability to read the full score and "hear" the sound of the notes in their minds. For others, scores are meaningless, being merely an odd jumble of lines and strange notations, with interpretive directions often written in a foreign language. Whether or not one can read scores, the act of listening to music as it is performed can be a wondrous experience. It is good to try to reflect on what a composer is seeking to say through the music as (re)created by the conductor and musicians. Some lovers of music enjoy dissecting a piece of music; for them, close analysis can lead to insight. Other people like simply to sit back, listen, and enjoy, content to let the conductor and musicians do the interpreting of the landscape of the score.

WHAT COMES NEXT

There are many routes to understanding and appreciating music. Seeking to understand landscapes in music is one route, and it is a fascinating one. My purpose is to illustrate the many ways in which composers have considered the central theme by concerning themselves with a variety of subthemes. I identify and discuss music from a variety of periods according to particular themes and subthemes rather than taking the standard approach, which discusses composers and their music according to accepted periods of time and style. The themes (chapter titles) and subthemes (section headings) were derived using a modification of grounded theory whereby I examined the huge amount of data and let the themes and subthemes emerge from that data rather than imposing my own organizational thoughts from the start.

The outline of the book is as follows:

- *Chapter 2* identifies how some composers have dealt with the passage of time.
- *Chapter 3* considers waterscapes as an extension of chapter 2.
- *Chapter 4* explores specific and generalized landscapes.
- *Chapter 5* deals with a variety of imagined and mythic landscapes, including some associated with nationalism.
- *Chapter 6* discusses music in which there is a search for meaning in difficult landscapes.
- *Chapter 7* focuses on death, survival, and remembrance, with subthemes including places of the dead, landscapes of war, and reactions.

- *Chapter 8* takes up a variety of subthemes related to music *in* places, including the existence of particular buildings, urban soundscapes, outdoor performances, and festivals.

This is not a history of music; nor is it a study in ethnomusicology, so any mention of regions and non-Western music is kept within the frame of the main concern. This study of landscapes in music ranges across time and space and considers a fair number of works by many composers from, especially, Europe, Russia, the United States, and, notably, contemporary composers from Canada, New Zealand, Japan, and China. However, it was not my intention to identify everything that composers of orchestral music have written over more than three centuries pertaining to the representation of landscapes in their music; rather, the music mentioned is representative.

The appendix presents a bibliographic essay on soundscapes and the geography of music in which the key works are identified by theme. The references include only those items directly quoted and cited.

2

Time and Space

TIME AND SPACE AS ESSENTIALS

The old saying that history deals with time and geography deals with space is not valid and never has been. Many historians acknowledge the role of space in societal development; all geographers are concerned with both time and space. The latter is so because of the need to discover and explain distributional patterns in space, spatial relationships and interactions, processes that cause change, and people's attachments to particular locations in space (places) as they are affected by associated processes (time). Time and space are inseparable, though some people stress time over space, while others reverse the order. On the inseparability, a Russian scholar writes, "Nothing, either in the world of physical objects or in the world of subjective ideas, exists in space without existing also in time or in time without also existing in space" (G. Orlov, quoted in Druskin 1983, p. 122).

Clearly, the concern for time and space links geography to music. Geographers use the concepts of space and time as they apply, for example, to how long it will take to go from here to there or to the meanings attributed to a portion of earth space by various groups of people at a particular time or over a period of time. Some composers do likewise, as will be demonstrated later in this book. Composers, musicians, and geographers also deal with absolute and relational time, but the perspectives taken in music involve a different orientation and scale from most geographical research.

What is the concern for time and space in music? Put simply, they are both essential! Music is set in time (the number of beats to the bar, such as

2/4 versus 6/8). Tempo markings written in the score or otherwise directed by the conductor identify the speed (or tempo) at which a passage is to be played (e.g., *adagio*, or at a slow pace, versus *presto*, or very quickly). As music is played, time passes bar by bar until, ultimately, it fills time (the length of the composition). Some music describes or manipulates time (as will be seen below), while other music is spatial inasmuch as it may seek to describe or represent the contents of a portion of the earth's surface (e.g., a national landscape), perhaps some spiritual space (e.g., a mountain), or a specific place (e.g., a spring on the side of a mountain). When music is performed, it is released in space, and for a second or two (depending on the reverberation in a particular space), it fills space (in a concert hall or an outdoor performance center) over time (the duration of the work) at certain tempi (the speeds at which the notes are played). Some music can evoke an image of space. For example, in his Symphony no. 9, Anton Bruckner creates a "sublime, cathedral-like vastness" by "the hollow D minor chord (*feierlich, misterioso*)," whereby the listener is taken from nothing to an astounding sense of immense space, a "*mysterium tremendum*" (Barford 1978, p. 61). Music played in time thus can create space.

From a different perspective, the placement of the musicians (in space) relative to each other can provide different sounds and varying senses of balance for the audience as well as the musicians (see chapter 8). For instance, it matters to the audience if the violas rather than the cellos are located to the front right of the stage (i.e., the stage as seen by the audience), for the dominant sounds reaching the audience will depend in part upon which of these string instruments is nearer to the front. Sometimes having players on risers (increasing in height the further back in the orchestra one goes) will alter the character and quality of sound reaching the audience. From a different perspective, some stages are difficult to play on, as when the players cannot hear each other well or, at times, at all, even though the overall balance of sound reaching the audience may be fine. Where the musicians sit can also have an impact on what they hear as they play. This can matter when, by way of an example, the cellos and the horns are to play a passage together but cannot hear each other. When this is the case, it may not be possible for them to feel in touch, so the conductor must keep the musicians playing together—in time. Another issue pertains to what sound is appropriate to a space, for the effect and impact of different instruments in different spaces matters. For instance, a solo violin or viola may be fine indoors but not on a football field (Maconie 1993, pp. 75–83).

SPATIAL-TEMPORAL LINKS

Some composers, being sensitive to natural elements and processes and to the impact on landscapes of human-environment interactions, may represent them in their compositions. In programmatic or descriptive music, we generally find that composers are concerned with both time and space (demonstrated in this and later chapters in many ways). Temporal concerns may be paramount, as when seasonal changes are the focus of attention. At other times, spatial processes may be especially important, such as with a river flowing down a hillside, waves smashing against a shore, or, to be different, a train traveling from here to there (see pp. 96–97). Sometimes spatial-temporal links are clearly evident, as with the hesitant movement of a stubborn mule (see p. 124), the awesome relentless advance of an army (pp. 155–56), seasonal changes in the countryside (below), or the vagaries of life in cities (pp. 85–95).

ODE TO THE JOYOUS CREATION OF THE EARTH

We first, logically, must consider the creation of the earth. One interpretation of Ludwig van Beethoven's Symphony no. 9 holds that it opens with the creation of the world. The uninhabited opening landscape (in the first movement), to my ear, is portrayed as a stark, cold, emotionless void. Later movements increasingly hint at the existence of humans, until, in the awesome final movement, Beethoven brings the hints together and develops them in a stunningly complete manner. The result is sublime, especially when the soloists and choir join the orchestra in the final movement. Among many other things, Friedrich von Schiller's words set to Beethoven's astounding music lead me—and many others—to ask why it is that we do not stand when the *Ode to Joy* is performed within the final movement. Oh, for the lack of a king! Why? By tradition, all audiences stand when the *Hallelujah Chorus* is heard. King George was so struck by the power and wonder of the words as set to George Frideric Handel's magnificent music that he stood up. Well, when a monarch stands, everyone present also stands, so the king's attendants and nobles stood with the king. The tradition of standing during the *Hallelujah Chorus* remains, even though the king is not present. But for now, the issue at hand is about time and space, not whether we stand or not, so let me continue with the theme, with some further thoughts about creation.

FROM CHAOS TO ORDER, GLORIOUSLY

When was the earth formed and when did nature in all its glory develop with humans as part of the scene? What was it like before there was human life? This is not the place to discuss evolution versus creationism, but the latter, relayed through a literal understanding of the Bible, led to the writing of one of the most remarkable moments in music, if not in all of the performing arts, namely, *Die Schöpfung* (*The Creation*), an oratorio (i.e., an extended sacred choral work intended for concert performance) by Franz Joseph Haydn (1732–1809). It explores the first six days of the "Creation." The first part is amazing for its representation of Chaos prior to life on earth. His thinking likely was influenced by extrabiblical thought. For example, Scottish scholar Donald Tovey surmised that Haydn would have read Immanuel Kant's thoughts on teleology (whereby he "proved" God's existence), published in 1755, two years before *The Creation*'s completion. No matter what Haydn considered by way of background, it is the work itself that matters in this case. The opening section was so amazingly ahead of its time in terms of how Haydn wrote his score that composers and others have long been drawn to it because of the remarkable orchestral effects achieved.

Haydn began his work with an orchestral description of a chaotic *Urwelt*, a formless void that only later took on form and structure, per God's wishes, and into which humans and other creatures were placed. In one of the first accounts of the work (included in Haskell 1996, p. 181), Hugo Wolf, a music critic in Vienna, noted, "The very first measures, with the muted violins, awoke in us the sensation of being in the presence of a mysterious something." The work's opening has dissonances, distorted dynamics, and various rhythmic values that cut against then-standard expectations. It puzzled listeners of the day for they were not sure what would happen next. Thus did Haydn create an image of disorder in a swirling void. But then comes the remarkable, crowning moment when order is revealed. The volume is *pianissimo* (pp, very soft) for the strings that accompany the barely whispered singing by the choir. Suddenly and unexpectedly, there is a stunning burst of gloriously bright sound, which disperses all suggestion of there ever having been but a void and chaos. The light "that pours forth through the music" (Tovey 1937, vol. 5, p. 116) lifts the listener from almost total stillness to a C-major full-textured chord played by loudly scrubbing strings, with mutes off, all woodwinds blowing, augmented by an extra flute and the deep growl of a bassoon, three triumphantly tooting trumpets, and the timpanist playing a loud roll. The critical, triumphant moment is reached as the word "light" is sung by the choir for the second time: *Es werde Licht, und es ward Licht* (Let there be light, and there was light). That awesome split-second transition is one of the most astounding moments in all of music. Haydn's genius is for all time, and, for all time, other composers have consulted his amazing work.

Haydn's music is stunning, no matter whether one accepts the creation story as having substance or as mere myth used to satisfy a human need for "a storied beginning" as part of a religious-cosmic order. In Haydn's work, nature was "literally" created in music and words, as were man and his female partner (a sexist order that was based on cultural norms applicable when the scriptures were written) and all other creatures on earth.

CREATION FROM DIFFERENT PERSPECTIVES

Interest in "creation" has been widespread, for tales pertaining to "the beginning" are to be found in the oral traditions of most societies (e.g., Campbell 1968; Tuan 1974). Composers other than just Haydn have tackled the issue. For example, more recently, the Finnish composer Uno Klami (1900–1961) wrote a work on *The Creation of the Earth*, which is the first of five sections of his *Kalevala Suite* (1943). He uses instruments not available to Haydn, including trombones, tuba, and readily retuneable timpani, and as a result, the work's feel is quite different from that by Haydn. Klami's chaos, if such it be, is agitated and mysterious in tone and color, starting from a fairly quiet opening, building gradually to a period of active vulcanism and seismic activity, with violent lifting and tearing of mountains. The listener thinks the work is to end with a grand climax, but the climax gives way to a long gentle passage, pastoral in character, with a quiet horn call, followed by a peaceful interchange between the strings, horns, and woodwinds. It is as if the first morning of life on earth has begun, and all in existence are called to live in harmony. The contrast between Haydn's and Klami's compositions is fascinating, but Haydn's work holds its own as the classic statement.

Yet another depiction of the creation of the world opens solemnly with the rumble of the lowest C of an organ undergirding a three-note introduction. The tone poem by Richard Strauss, *Also sprach Zarathustra* (1896), opens with a trumpet declaration that is followed by two thunderous notes of full brass with the double basses then playing a long, wonderfully resonant note, until a timpani roll swells into a solo of thirteen beats on alternating notes. On the final note, the trumpets restate the first three notes, and everything else is repeated. On the third go round, there is a crescendo into a brash, full-orchestra phrase and a gigantic concluding chord in which an organ and crashing cymbals join. In the midst of this astonishing climax, the orchestra suddenly stops playing, leaving only the organ playing an eerie C-major chord. In those few dramatic opening phrases, Strauss takes us from nothing to the creation of the world. The strings and woodwinds then sing a solemn ecclesiastical theme to announce the existence of humans, a theme that is developed later in the work, and so, the music proceeds. This dramatic musical description of the creation of the world was used as the music for the opening scenes of the (then) futuristic movie *2001*. Strauss would have been surprised and likely pleased to see how his music was used in the movie.

Given these several references to creation, we will now assume that the earth is no longer devoid of presence and meaning. Let us further assume that, with the "birth" of light, substance, order, and human, animal, and other life on earth, "voidness" has given way to *space*, as per the works by Haydn, Klami, and Strauss, and that, therefore, simultaneously, *time* also has come into existence. As discussed earlier, anything in music automatically involves time, but is time inevitable? George Benjamin (1960–) in *Sudden Time* (1993) considers a world without time but within which time suddenly starts. He also considered time in *Upon Silence* (1990). The two works consider stretchy dreamtime and regularly moving real time. The contrasts between these two forms of time resonate with audiences given that, periodically, we all have varying senses of time, sometimes vague, at other times highly structured and pressured. In his music, Benjamin expresses time as condensed and, yet, also extended, as when his richly original orchestral soundscapes are periodically disrupted by "pulses" that are in opposition to each other. It is worth noting that Benjamin was inspired to write as he did by the work of poets: *Sudden Time* was inspired by a poem by W. B. Yeats, whereas *Upon Silence* refers to a poem by Wallace Stevens.

DREAMTIME

Most readers of this book likely think of time from a Western perspective, with due regard for relative and absolute time. The development of the digital clock changed our concept of time: with the analog clock or watch face, we think in relative terms ("it is about one fifteen"), whereas with the digital clock, we say authoritatively, "It is one seventeen." Stating that it is 1:17 suggests that the timepiece is accurate and, thus, correct, no matter whether or not such is the truth. But not everyone thinks of time in this absolutist fashion. Australian Aborigines, for example, have a concept of time that is at total variance with Western notions.

Dreamtime, or the Dreaming (but not as I would dream in my sleep), incorporates eternal, creative, and ancestral time that is simultaneously spatial and temporal, each with many layers and filled with myths (Berndt and Berndt 1964, pp. 187–88). To put it another way, Dreamtime, or the Dreaming, refers to the period and the spirit world before time began, when the land was formed and the spirits roamed free. Contact with the Dreaming is critical to the continuance of Australian indigenous identity. Even urban Aboriginals in Australia today, who may feel alienated from their traditional ways in many respects, are generally still able to relate to and depend upon their attachment to the Dreaming, which, in turn, links them spiritually to a critical aspect of identity that burns within them, something that is beyond the full grasp of Europeans.

Some Australian composers have sought to incorporate into their work the Aboriginal concept of Dreamtime. For example, Richard Charlton's guitar suite, *Impressions from the Dreamtime* (1988), tries to convey the feeling of such time, drawing, in turn, upon Aboriginal notions of the legend of fire, a frog dreaming, a corroboree, a black swan, and a finale in which dancers are changed into animals. The imagery Charlton draws upon is all Aboriginal. A different approach has been taken by American Steve Roach in *Dreamtime Return* (1988), in which he ties recordings made in various Australian locations with synthesizers and percussion. He seeks to activate listeners to reach a deep level of consciousness that draws upon the trance-inducing music he has created.

Australia's leading composer, Peter Sculthorpe (1929–), has written several works that seek to address the elusive concept. In *Jabiru Dreaming*, String Quartet no. 11 (1991), for example, he uses the term *songlines* to imply Aboriginal traditions of associating songs with geographical features and (again, jointly spatial and temporal) paths of ancestors. He includes birdsongs (such as those of Blackbirds) and insects chirping and clicking in his music for he believes they are essential Australian sounds. In *Kakadu* (1988) for orchestra, he draws upon an Aboriginal song set in the location of the *Crocodile Dundee* movies (i.e., the Kakadu National Park), within which are places sacred to Aborigines. In *Kakadu*, the strings produce bird sounds, bongo drums provide a steady rhythm, and the brass play a series of accented chords that lead to a descending melody on the trombones and then horns. Other works include *Into Dreaming* and *Dream Tracks*. These several examples of Westerners trying to understand and represent a non-Western construct of time and space are fascinating; yet, it may not be possible to truly grasp these Aboriginal constructs in terms other than their own. At this time, there is no readily identifiable Australian Aboriginal composer of orchestral music.

TIME TRIUMPHANT

Most readers of this book will find it hard to conceive of Dreamtime. It is easier to imagine and use the two most basic Western methods for measuring—and keeping track of—time, the diurnal and seasonal methods. First, the workday for some is decided by the changing placement of the sun relative to our east-west horizons. Reading the sun's movements and noting its seasonal adjustments is an ancient skill applied and used for survival. For most people today, guesswork or the sundial or other gauge has been replaced by the clock (and watch).

Haydn's Symphony no. 101 (*The Clock*) takes its name from the regular "ticking" of the basses in the second (or *andante*) movement. Haydn's clock of 1794 was somewhat akin to my father's clock. I have childhood memories of the regular ceremony of clock winding that only he could perform. Haydn's "clock movement" opens with strings and the bassoon, "tick-tocking" in a steady but rather monotonous beat that serves as the accompaniment to various melodies played as the movement proceeds. Some lovely color is added to that beat by the use of other instruments at various times during the movement.

Another orchestral clock appears in the "Viennese Musical Clock," part of the *Háry János Suite* (1927) by Hungarian Zoltán Kodály (1882–1967). The famous clock in the imperial palace was renowned for its musical discourse as brightly painted models of soldiers "marched" in time to the music. The constancy of the march is at one and the same time the regular beating of the clock. At the end of the light and sparkling piece, bells join in for a joyous climax.

TIME RELENTLESS

Can time stand still, as the saying goes, or can it be suspended? Here, I am not referring to a clock whose spring runs down or that stops when the battery dies. Instead, I refer to the philosophical issue of whether or not time can actually stand still or be suspended. The English composer Harrison Birtwhistle (1934–) clearly believes it can at least be nearly suspended, for in *Earth Dances* (1986) the "blocks" of music respond as do the earth's continents; they are in perpetual, yet imperceptible, motion. Birtwhistle's score has the orchestra divided into six sections, as if they were separate geological strata, shifting and shuffling around each other in ever-changing patterns, eventually returning to their starting points only to find that the view has changed; so, the cycle starts again. Birtwhistle has described his works as "imaginary landscapes." This phrase is most apt with respect his *Triumph of Time* (1972), scored for a huge orchestra. The work can be likened to moving through a landscape but seeing the same features from ever-changing perspectives rather than always seeing new features. Time constantly moves and changes, as demonstrated within this work. Indeed, time simply cannot be stopped. It is relentless.

Birtwhistle was inspired to write *Triumph of Time* after seeing the sixteenth-century Flemish artist Pieter Breughel the Elder's troubling painting of time as a remorseless force. Breughel shows time as a figure in a cart leading a procession that moves ever forward, crushing humankind in its inexorable march. Only cyclical events like tides and seasons, evident in the background, remain untouched. Breughel's painting is deeply troubling for many viewers, and Birtwhistle's musical landscape is deeply troubling for some listeners. Indeed, some listeners cannot go beyond the at-times raucous and dissident work as it constructs the coexisting times of the past, present, and future. Whenever the work is performed, audiences tend to be polarized between those who rejoice in having heard a wondrous work and those who are thankful it is over! With patience, and at least a second hearing, the keen listener becomes aware of both stunning subtlety and immense power in the way the work conveys the constant passage of time. To my mind, Birtwhistle's jagged musical landscape of time and place, seen from different angles, is brilliant. Perhaps the dichotomous audience reactions to the performance of *Triumph of Time* reflect an essential truth about orientations at the turn of the twenty-first century, namely, those who looked to the past for comfort and those who looked ahead to new challenges.

CYCLES OF TIME: MONTHS AND SEASONS

Canadian composer Gilles Tremblay's (1932–1982) *Solstices* (*The Days and Seasons Revolve*) (1971) for chamber orchestra considers cycles of time based on the months of the year and also the turning of the four seasons, all interwoven. His score has twelve "fields" paralleling the months of the year. While each of the fields (or months) is distinct, the seasons are represented by different instruments (spring, flute—bright and lively; summer, clarinet and percussion—insect sounds; autumn, double bass—varyingly stormy and silent; winter, a horn—mournful, with silence and slow evolution). Some of the fields are linked by periods of silence, while others are linked by full-orchestral sounds. Fascinatingly, each performance is unique because Tremblay asks that the month in which the music is performed be the point in the score at which the musicians begin to play. Bruce Mather (2001, p. 716) observes that the "interplay between the instruments is principally dependent on 'réflex,'" whereby one player reacts to the playing of another.

Seasonal changes are often celebrated by people close to the land, no matter what culture they are from. For example, all over the world, there are rituals of thanksgiving pertaining, especially, to harvest time. Thus, for example, there is the Christian Harvest Thanksgiving celebration with special hymns in European and European-derived societies (including Canada, Australia, and New Zealand) and a secularized Thanksgiving in Canada (in October) and the United States (in November), with attendant rituals such as turkey carving and cranberry sauce as part of the large, family-centered meal. Other rituals apply in many cultures to the time around planting. As a side note, Vermonters have yet to find a composer willing to compose something on Vermont's fifth season—mud season—which occurs between winter and spring!

If we desire a happy impression of winter, Pyotr Ilich Tchaikovsky's Symphony no. 1, *Winter Dreams* (1866), fits the bill. The first two movements have programmatic subtitles—"Dreams on a Winter Road" and "Land of Gloom, Land of Mists"—whereas the third and fourth movements have none. Tchaikovsky composed the work while, he said, he was an "immature" young man. He may still have been searching for his "voice," but the voice we know from his later, mature symphonies was clearly well on its way to fruition within this work. Criticized by his grouchy former teacher, Anton Rubenstein, the work nevertheless is well worth listening to because of its charm, lyrical folk song–like tunes, and innovative use of timpani at the end of the third movement. Seemingly, only two themes within the work have any resemblance to actual folk songs.

Sleigh bells are a favorite musical geography signal that winter has arrived. Numerous composers use them to convey a sense of time and color in winter, including the American composer Leroy Anderson and his famous *Sleigh Ride* (1948), with rhythmic jingle bells that ring in time with the prancing horses. The work is played ad nauseam during the Christmas period in North America, as often by school orchestras and bands as by closeted elevator-music operators in far too many stores. A lesser-known work in a winter setting was written by the Englishman Frederick Delius (1862–1934) during an 1887 visit he made to the home in Norway of composer Edvard Grieg. Delius discovered that Grieg's guests were expected to contribute to the evening's postdinner fun, so he sat down and wrote a piano piece about a sleigh ride in which the horses pull the sleigh, varyingly galloping, walking quietly, galloping again, and then walking quietly to the destination. What a gift to the next evening's soiree! The work was later scored for orchestra.

A thirteenth-century song reminds us that "Sumer Is Icumen In," and once it arrives, the most exquisite celebration of summer undoubtedly is Ralph Vaughan Williams' *The Lark Ascending* (1920). The orchestra's three-bar introduction evokes an image of a hot, languid summer's afternoon, and then, in the fourth bar, the lark's delicate song emerges, played by solo violin. The purity of sound is magic. The violin music says, in essence, sit back, relax, and let me pour over you! The orchestra continues to provide quiet backing to the pleasant beauty of the violin's simple and effective evocation of sweet-smelling flowers and incessantly humming insects.

Prélude à l'après-midi d'un faune by Claude Debussy (1862–1918), suggested by a poem by Stéphane Mallarmé, is concerned with the longings of a faun as he watches some nymphs on a hot summer afternoon. The solo flute's melody first conveys an atmosphere of haze during the heat of the day and the faun's yearning. The languorous melody is picked up by the horns, backed with strings and harp. The flute returns, and the gentle mood continues from there, as various turns are taken by the orchestra and harp, sometimes with clarinets dominant but with the flute almost always present, ever so sweetly conveying the sense of the faun's wonder and longing. This lovely musical landscape is now one of the most popular of all of Debussy's many impressionist works. The sense of a magical summer landscape is complete, and in this work, time almost stands still, so mesmerized is the listener by the score.

THE YEARLY CYCLE OF SEASONS

Why focus on just one season of the year? So thought Antonio Vivaldi (1678–1741). Why not draw on bird songs and bothersome bugs in spring, summer thunderstorms, autumn harvest celebrations, and bitter winter winds when representing all four seasons? Vivaldi wrote of the seasons of the year and associated phenomenon in nature in his *Le quattro stagione* (*The Four Seasons*). It is an early masterpiece of pictorial music that was popular in the Baroque period. As such, its musical structure and sense of time is utterly different from that explored in Birtwhistle's twentieth-century work; of course, Birtwhistle and all others discussed in this chapter likely could not have composed what they did if Vivaldi and, later, Haydn and their contemporaries had not explored their creative art to the limits that they did.

Vivaldi was a pioneer in the descriptive form of composition. His prime goal was to describe in music as literally as possible personal reactions to the aesthetic of nature. Unlike the freedoms later to be appreciated by composers during the Romantic period, Vivaldi and his compatriots felt bound by the existing constructs of musical order, so each concerto typically had fast/slow/fast movements, and the musical quotes were as literal as they could be made. Composed between 1720 and 1725, Vivaldi's twelve concertos for violin and string orchestra with continuo (harpsichord) are wondrous, but four of them are played most often, perhaps due to their collective title, *The Four Seasons*. Each of the seasons is based on a poem (which he had preface each composition). The orchestra opens each concerto, and then one or more violinists "tone-paint" an array of descriptions and feelings associated with an Italian landscape.

"Spring" opens with gentle breezes blowing, birds happily singing, and streams flowing gently. The sense of bliss is shattered by a short, sharp storm, with lightening and thunder (wonderful "scrubbing" by the string players), but the storm soon subsides, and the birds sing again. The second movement suggests leaves are rustling, while a shepherd sleeps (solo violin) and his dogs bark (an insistent rhythm on violas). The final movement is a shepherds' dance, the music cleverly droning, suggestive of bagpipes. A sense of sluggishness opens "Summer," but then three violins sing (using high pitches and trills) to represent a cuckoo, a turtledove, and a goldfinch. A sense of foreboding is created when the music takes on a sharp edge and gets louder. A great section follows, with a swarm of buzzing insects represented by the strings rapidly and repeatedly playing low notes. Mosquitoes? Bees? Wasps? We do not know which, but the representation of the little creatures is so skillful that the listener can be quite distracted, even to the point of feeling ill at ease. Another storm follows. In "Autumn," peasants dance in celebration of their good harvest. The people get drunk (solo violin) and fall asleep (a quiet cello line). Next morning, they go hunting. A wild boar is terror stricken (triplets on the violin) and seeks to escape (whole orchestra). "Winter" follows. We shudder at its suggestion of the cold wind (solo violin). Walking on ice is a slow task; someone falls (quick descending runs on the strings) and is helped up; sounds of the ice cracking and a bitter wind end the work. *The Four Seasons* delighted Vivaldi's contemporaries and still delights audiences today. The point here is that with skillful scoring, Vivaldi got his orchestra of string players to create all sorts of sounds in nature that were fully recognizable to his audiences. And that is the point. The sounds were readily understood. Vivaldi and his contemporaries created musical geographies, so intentionally vivid were their references to nature, all caught within the constraints of musical composition of the time.

Descriptive writing in music changed as composers developed new insights concerning how to use harmony and timbre (tone color), new types of rhythms, and new structural forms that would surely have surprised seventeenth-century and eighteenth-century composers and their audiences. Orchestras increased in size, and players were forever playing "new" music. By the mid-nineteenth century and on into the twentieth century, many composers had become skilled at making programmatic and descriptive musicscapes. Excellence of playing is required for a good performance of Haydn's work, as is true of the music of later composers. But new skills were demanded as time went by; thus, players in due course came to contend with such remarkable silliness as Nikolay Andreyevich Rimsky-Korsakov's *The Flight of the Bumblebee* (which lasts for about eighty seconds, when played by professionals) and a much earlier, somewhat similar (but slightly shorter) work by Franz Peter Schubert called *The Bee*. Each is a trial by fire for players, but a treat for audiences. In these pieces, time literally flies!

British composer Birtwhistle also wrote an opera, *Gawain* (1991), in which the hero's trials are set against the remorseless cycle of the seasons. His sense of the passage of the seasons lacks the charm of Vivaldi.

And, again differently, in Gustav Mahler's 1909 *Das Lied von der Erde* (a song cycle on *The Song of the Earth*), concern for nature's ever-renewing processes identifies the transience of the human spirit, for while we die "the dear earth everywhere/Blossoms in spring and grows green again!/ Everywhere and eternally the distance shines with a blue light!/ Eternally . . . eternally."

In another vein, in 1970 Tōru Takemitsu wrote *Shiki* (*Seasons*) for four percussionists performing on pitched metallic percussion. Takemitsu had a close association with the skilled players of Nexus, the Canadian percussion group that performed this and many of his other percussion compositions. Another of his works, *From Me Flows What You Call Time* (1990) for Nexus and orchestra, evokes the passage of time in terms of representing the now more than a century of musical groups that have passed through Carnegie Hall in New York. A sense of flow is captured in the imagery of the Tibetan "Wind Horse" that Takemitsu had in mind as he composed the work. Drawing upon Tibetan beliefs, he used the colors (represented by ribbons in the performance space) associated with the horse: green for wind, blue for water, yellow for earth, red for fire, and white for the sky.

Russian composer Alexander Glazunov (1865–1936) also tackled the four seasons of the year. *The Seasons* (1900) was written for the Russian Imperial Ballet; hence, it has a different character than the works identified above. The seasons are of varying length, and each season has several sections. The four of sections of "Winter" are especially descriptive. First, we can imagine looking out the window and seeing frost (a polonaise) sparkling in bright sunlight. Then, comes ice (a dance played by violas and clarinets), which conjures up the image of someone gracefully skating across a frozen pond, happy with the beauty of the moment. Next is a scherzo that creates an image of stinging hailstones, conveyed by staccato strings and winds with menacingly long-bowed notes on the low strings. Finally, snow (a waltz) has a gently moving quality, conveyed by solo oboe and trombone, as we imagine the wonder of big, soft snowflakes floating down from the sky and settling on tree branches and all else that will accept ever-larger little piles of snow. "Spring" opens with a harp, and we then hear various birds and a representation of flowers and a gently blowing breeze. "Summer" is said to be set in a cornfield: cornflowers dance, but the heat exhausts them, and a bird, too, is left exhausted. Frustrated by the constancy of the breeze, a frenzied bacchanal introduces "Autumn," which includes brief hints of winter, spring, the bird, and the wind, as if reminiscing about the year that is ending. A quiet waltz replaces the bacchanalian scene, but the bacchanal returns once more as leaves fall from the trees. The work ends in darkness, with, in the final statement, reminders of the four seasons of the year.

SPRING: SEASON OF NEW BIRTH

The season to attract most attention from composers undoubtedly is spring. In *Green* (1968) Takemitsu created a very gentle, peaceful work. He was inspired to write the work as an expression of the foliation of spring with the transforming of buds into young leaves. This lovely work was written for his daughter and the daughters of friends.

Not all music that refers to nature's seasons is programmatic. An excellent example of this point was created by Robert Schumann (1810–1856). Schumann had a remarkable outpouring of music when, following his marriage to Clara Wieck (a virtuoso pianist) in Vienna in 1840, he wrote nearly 150 love songs for her and, in just four days early in 1841, sketched out his Symphony no. 1. Inspired by Adolf Böttger's poem about spring and by his own *liebesfrühling*, Schumann called the work his *Spring Symphony*, but he claimed that neither any painting nor the poem was active in his mind when he composed; indeed, he said his desire was to avoid purely descriptive music. Nevertheless, he gave the following titles to the four movements: "Spring's Awakening," "Evening," "Merry Play," and "Spring's Farewell." He subsequently withdrew the titles. The work is neither fully descriptive of the landscape nor of nature as something external. It is a delightful nonprogrammatic exploration of new life, light, and happiness and, especially in the final movement, of, yes, the triumph of spring.

The British composer Frank Bridge (1879–1941) opened his *Enter Spring* (1927) with a flurry of sound, possibly to represent an urgent need to get rid of winter, before a serene image of a lovely pastoral setting takes over and lulls the audience into a sense of peace. Thirteen years earlier, he had written *Summer*, also rather idyllic, composed in the style of Delius. This work contrasts with the "sombre plangency" of his 1928 work, *There Is a Willow Grows Aslanta Brook* (Paine 2001, p. 347).

U.S. orchestral audiences regularly hear Aaron Copland's *Appalachian Spring* (1945) in concert performances. The chamber orchestra score was extracted from the original ballet score of the same name, but the work is now commonly heard when played by a full symphony orchestra. The title was taken by Martha Graham from a poem by Hart Crane. Interestingly, the poem's title, *Frühling in Pennsylvanien*, should have been translated as *Spring in Pennsylvania*, but the name *Appalachian Spring* is now firmly attached to the work. The best-known theme in the work undoubtedly is the Shaker dance-prayer, *The Gift to Be Simple* ("'Tis the gift to be simple/'Tis the gift to be free/'Tis the gift to come down/Where we ought to be"). This beautiful song is presented by the solo clarinet and then explored in five variations. Copland's compositional style was outwardly plain and simple (akin to Shaker belief), including, as it did, long phrasing (e.g., on the violins) and a mix of time signatures. William Austin (1986, p. 500) has called Copland's sound "never quite conventional, nor complacent in its novelty." The review in New York's *Herald Tribune* (May 15, 1945) stated, among other things, that the work concerns "a pioneer celebration in spring, around a newly-built farmhouse in the Pennsylvanian hills in the early part of the last century." The young couple is given advice by various people about how they should live their new life together, but in the end, they realize that they must make their own decisions. As the work comes to a close, it is easy to think of a pleasant spring evening, with the setting sun streaming its last light through the fresh green leaves of the trees on the hillside. The pure joy and delight with life in this work by Copland differ from the next work.

ASTONISHING NEW BIRTH

The intensity of "new birth" in spring is especially wondrous to see in those environments in which there is a long, hard winter. Buds on trees suddenly appear and then burst forth in tender greens that are all too soon lost as more mature greens emerge. The idea of new birth coming from the earth, from nature, caught the imagination of a Russian composer who created one of the most astonishingly rhythmic works of all time. At its premier performance in Paris in 1913, however, many members of the audience were so shocked by the music that a riot occurred. No matter, it forever changed the nature of music. Just as Haydn's *The Creation* was revolutionary in its way, so too was Igor Stravinsky's (1882–1971) composition in honor of spring.

Stravinsky wrote *The Rite of Spring* (1913) in a manner that surely would have shocked Haydn's audiences and likely Haydn himself. Called the enfant terrible of contemporary music, in part due to *L'oiseau de feu* (*The Firebird*) and *Petrushka* ballet scores, written for the legendary ballet master Sergei Diaghilev, Stravinsky accepted Diaghilev's invitation to write a third work. The result was to mark the beginning of modern music. The work was based on a dream Stravinsky had and then related to Diaghilev. *The Rite of Spring*, subtitled *Pictures of Pagan Russia in Two Parts*, is often referred to by its French title, *Le sacre du printemps*. Interestingly, Stravinsky's Russian title for the work translates into English as *The Coronation of Spring*.

It was not Stravinsky's title that was controversial but the work itself. The music was the complete antithesis of all music previously written about spring. Whereas previous music about spring was generally chirpy, in more ways than one, his music had a relentlessly savage quality in its portrayal of, in Stravinsky's words, "the surge of spring, the magnificent upsurge of nature reborn." In *The Rite of Spring*, spring is seen from within, writhing in the pain of labor, and then being delivered. The dance sequence reaches a climax as a maiden is sacrificed to nature by dancing herself to death. Each of the work's two parts, "The Adoration of the Earth" and "The Sacrifice," have a number of sections. So varied are the tempi that many in the first audience thought the work chaotic, perhaps because some of the musicians had trouble playing the complex rhythms and ever-changing pulses that were quite new to them and difficult for many to comprehend. The same held for the dancers, who had to perform the series of rituals held in propitiation of nature just prior to the arrival of spring. Indeed, they had a tough task dancing to music that was strikingly different from anything ever heard before. Even the lead male dancer, the great Vaslav Nijinsky, the leading dancer of the time, needed 120 rehearsals. Following the memorable opening of the ballet, Stravinsky wrote a concert version, without dancers, which was first performed in 1914, and it is now played around the globe.

According to Finnish conductor Jukka-Pekka Saraste, Stravinsky wanted the bassoon to sound earthy and uneasy, like something coming out of the earth. The French bassoon of the early part of the twentieth century had a much more nasal sound than the smoother sound of today's bassoons, so it is harder today to get the sound Stravinsky wanted. *The Right of Spring* ranges in color from dark, earthy, and raspy to vivid, bright, and wondrously imaginative. There are many short, simple, folklike melodies, but, essentially, the work is rhythmic, not melodic. The spring landscape created by Stravinsky, with the image of dancing figures easily visualized as the vivid music proceeds, signals an earth-awakening spring ritual like no other.

NEW BIRTHS ELSEWHERE

Stravinsky's *Sacre* has influenced other composers. For example, Finnish composer Uno Klami's five-part *Kalevala Suite* uses Maurice Ravel's "slick and sophisticated orchestral technique," Stravinsky's "modern primitivism," and his own harmonic cadences in, especially, part five, "The Forging of the Sampo," the result being a stunning "Finnish *Sacre*" (Salmenhaara 1994, p. 6).

New Zealand's Gillian Whitehead (1941–) was inspired by Stravinsky's remarkable score to write *Resurgences*, a work William Dart (1995, p. 8) calls "an Antipodean *Rite of Spring*." *Resurgences* captures the sense of new life akin to the gigantic burst of energy in spring, only in this case it is the eruption of energy from beneath the ground. Whitehead's work concerns itself with a geyser's periodic explosion of life, inspired by the geothermal activity in the vicinity of the city of Rotorua, inland North Island, and, as she states, "living away from the sea and being drawn back to ideas of the sea, ideas which are very strong in all New Zealanders—looking out to distant horizons" (quoted in Dart 1995, p. 6). Whitehead's work opens in an evocative manner, with snippets of sounds followed by silences that promise more to come. It is as if the underground forces are bit by bit getting ready for a new display of energy. Tension and volume gradually build as the snippets take on a form that is held together by orchestral rhythms dominated by percussion. The eruption provides a fascinating musical display of power and vivid design (read also about the geyser by Jón Leifs in chapter 6). The dramatic interplay of rhythms and musical themes representing the changing character of the "blow" abruptly halts as a mysterious element emerges from the woodwinds and brass, but the percussive rhythms cannot be held back for long, and the energetic drive returns, only to be replaced by a rude clashing of forces represented by driving percussion and stuttering brass. The eruption has concluded, it seems, and a time of peace returns, with a bassoon and strings relishing their roles as intermediaries between outbursts. The dramatic playfulness of the soundscape reminds the listener that even when far away, the essential character of nature is ever present. Energy spent, the work comes quietly to a close. Just as Stravinsky makes use of stunning percussive rhythms, so Whitehead's drama revolves around marvelous percussive rhythms that brilliantly capture the burst of energy from the earth. Her use of a geyser to symbolize resurgence in nature leads us to "see" in a new and sharper light other forms of short, sharp "energy expenditures," as with, on a different scale, buds bursting into life on tree branches.

Another "first stirrings" is expressed in *Landscape* (1988) by the English composer Diana Burrell (1948–). The work opens with birdlike (bassoon and cor anglais) keening calls that transport listeners to a weird, primeval landscape with various and curious percussive instruments. A magic moment comes when two tenor recorders "emerge as the first stirring of life from the bowels of the earth to which the music has sunk" (Fuller 1994, p. 82). Without doubt, capturing the wonder of spring has been a special challenge to many composers, most of whom, especially following Stravinsky, accepted that spring is a violent time, though perhaps this is a matter of scale, for from afar, a flower pushing up out of the ground may seem to involve an easy process, but up close it is possible to see the energy needed for this miracle to occur.

TIME AND MOVEMENT IN THE COUNTRYSIDE

Time, here, has a double meaning: time spent in that part of space we designate as countryside and time that passes there as things change. Movement, here, has a triple meaning: it pertains to all aspects of the work, including itself (the passage of time from the start to the conclusion, time markings and tempo changes in the score, and so forth), to the processes represented (such as movement into and around the countryside), and to natural forces (such as the wind and rain).

Ever since "God created the heavens and the earth" with human inhabitants as per Haydn's *Creation*, people have wandered through numerous landscapes. The idea of wandering through landscapes did not start with the notion of relaxation. For centuries, wandering was related to basic survival, as in prehistoric hunting-and-gathering societies. A considerable time later, wandering—purposeful wandering—was generally undertaken by those who had wares or a skill to sell as they went from town to town. And then, there were those for whom the adventure of exploring new lands was meaningful. Our concern is with early nineteenth-century wandering, when hiking with vigor or more gently strolling through landscapes came to be a leisurely activity for many in Western societies, especially those with wealth and the gift of time. These wanderers sought what the countryside could bring them: relaxation, enjoyment, and inspiration.

As the industrial revolution expanded the middle classes and trains provided Sunday outings for the working classes, "wandering" became increasingly popular, if only for one day a week. Stanley Sadie has observed, "Natural rural beauty was only just beginning to be admired; the previous generation, the age of landscape gardening, had held that only unruly nature could be bettered by human art" (Sadie 1990, p. 271). By way of illustration of this point, there was a long tradition of "tramping" in, for example, Germany, where geographers and others enjoyed outings, and Czechoslovakia, where people were taken by stories of the American West and Karl May's crazy "Western" books (he never saw the West). Beethoven loved to wander in the countryside near to where he was staying (see p. 74) and Felix Mendelssohn wandered in parts of Britain and Italy (see pp. 64, 77–79).

French composer Hector Berlioz (1803–1869) was especially struck by what he saw and experienced in the landscapes he walked and rode through while visiting Italy. He thereafter, in 1834, composed a wonderful landscape in music called *Harold in Italy, A Symphony for Viola and Orchestra*. In the work, Berlioz creates a wonderful dialog between the violist and the orchestra as Harold traverses the Abruzzi landscape near Rome. The impetus for composing the work is interesting. On a trip by boat from Marseilles to Leghorn (Livorno), Italy, a sea captain told him of having ferried Lord Byron around the Greek Isles. Berlioz recalled that Byron had written *Childe Harold's Pilgrimage*. About the same time, Niccolò Paganini, the famous string player, asked Berlioz to compose a viola concerto, so Berlioz put these two matters together and, over a three-year period, wrote *Harold in Italy*. Berlioz's notes tell us he drew upon his impressions from his recollections of his wanderings in the Abruzzi region. He also wrote that Harold is a sort of melancholy dreamer, in the style of Byron's Childe Harold; hence, the title of the work. Paganini was not pleased with the result, however, because he wanted a concerto to show off his virtuosity, whereas the composition had the viola and orchestra working together in such a manner that it has been called a symphony with an obbligato viola (that is, where the viola is essential to the composition but is secondary in importance to the principal melody). Another player gave the premier performance.

The first of Berlioz's melodies played by the violist is an idée fixe (Harold's theme), which is reproduced and developed in each of the four movements. There is little connection between the music and Byron's poem other than a strong romantic feeling and the sense of a person in movement. One quickly gets the impression that Harold is rather laid back due to the music's gentle and pleasing manner. (There is no bizarre quality to this music as there is in *Symphonie fantastique*, also by Berlioz.) The landscape is a major part of this musical portrayal. Clever uses of instruments convey landscape sounds. For example, an inventive use of two harp notes doubled by the flutes, oboes, and horns creates lovely convent bells that ring periodically as we hear Harold walking along at a leisurely pace. At the outset, Harold climbs in the mountains where he hears an Abruzzi mountaineer serenading his mistress. Later, he watches a march of pilgrims singing their evening song, sees and, perhaps, even participates in a brilliant tarantell-alike dance that tells us of a brigands' orgy, and, to conclude the work, recalls memories (in which musical reference is made to themes stated earlier in the work). At its simplest, the music is about the passage of time in a specific space, as Harold wanders from place to place and witnesses life in a beautiful cultural landscape. But, of course, Berlioz creates something considerably more interesting than simply a depiction of the passage of time in space.

Takemitsu provides a quite different approach to time and movement in landscapes. His scale is more intimate and, thus, he was able to consider details not possible for Berlioz. Using the image of a Japanese garden, in *Quatrain* (1975) for clarinet, violin, cello, piano, and orchestra, Takemitsu portrayed "a picture being unscrolled. . . . The scenes change successively without a break. In other words, it is similar to the relationship of a garden and a person walking to it" (cited in Narazaki 2001, p. 23). The "unscrolled" garden reveals delightful detail in magical colors. Takemitsu also used the image of a Japanese garden in his *Tori wa hoshigata no niwa ni oriru* (*A Flock Descends into the Pentagonal Gardens*) (1977). In that score, the garden is approached from a distance. The flock of birds (principally the solo oboe, with "sharply contoured melodic figures") flies around the garden, which, Yoko Narazaki (2001, p. 23) observes, is symbolized by a "static succession of harmonic fields." In *Garden Rain* (1974) there are slow moving, widely spaced chords. Takemitsu's garden landscapes are a mix of static form and gentle movement on a very personal scale. It is hard to imagine his scenes moving on, as it were, to the next season.

Delius also took great delight in his garden, in the manner of many of his English countryfolk. *The Walk of the Paradise Garden* (1908) is "gently romantic music" that with tenderness and graciousness "puts colors and fragrances into sound" (O'Connell 1950, p. 141) in a manner reminiscent of Debussy's tone painting.

Whereas Takemitsu and Delius focused their thoughts on details within such small areas as gardens, the English composer Arnold Bax (1883–1953) took a wider look at landscapes around him, including beech woods near his house in the Chilterns, where he and his brother's family enjoyed walking. He wrote three woodland-inspired compositions, including *November Woods* (1917). For a large orchestra, the work contains subtle color changes that may mirror Bax's appreciation for what he saw when was in the woods: the music suggests that only slivers of light came down to him at certain times, whereas at other times, winds swept the branches and leaves from side to side, and the contrasts of light were bold. His *Summer Music* (1920) is different, for it captures a woodland in southern England at noon on a hot, windless June day. Especially beautiful are the opening horn solo and the cor anglais, which takes over. These two works capture moments in time: one at the cusp of seasonal change, the other at the height of one season.

From a different tack, we can note some of Johann Sebastian Bach's work. Amid his huge outpouring of compositions are four Suites for Orchestra (each with several short pieces), which are delightful for the evocation of charming scenes of dancers. We can imagine that some of the scenes are filled with children; their elders dance a saraband (of Spanish origin) in one scene, a minuet in another, and a gavotte (a dance once favored by French peasants) in a third. These works are filled with movement.

CONCLUSION

Movement in time and space is fundamental to music and to geographical analysis. If *The Creation* best captures the notion of chaos and the sudden change to light and life—the start of time and space, if you will—then *The Triumph of Time* best captures the relentlessness of time, and *The Rite of*

Spring best captures the rapture and wonder of the earth's recovery from winter, of new creation. *Harold in Italy*, on the other hand, delightfully expresses the leisurely exploration of a particular cultural landscape—part of the earth's space—as time passes, measured by the tolling of bells and the unhurried walking pace. In contrast to the purposeful pace of Harold's exploratory walks, *The Rite of Spring* and also *Resurgences* have an urgency to them, as if nature simply cannot be held back. Nature *must* emerge—*now*! Physical and human processes are constantly at work; therefore, anyone seeking to understand the importance of change must determine the speed of the processes causing the change. Some processes are more rapid in their shaping of the surface of the earth than are others.

In the works discussed in this chapter, the link between landscapes (as an aspect of space) in music and the passage of time, has been quite explicit. The same will pertain in many of the works to be discussed in later chapters, though reminders of this fact will not always be presented. Time and space and movement are also explicit in the next chapter, which focuses on water.

Before turning to waterscapes, we can conclude from the material presented in this chapter that many composers are skilled at "reading" landscapes and representing in their music something about those landscapes, with due awareness of time and space. Does this mean that composers are geographers? Not necessarily, but it is clear that many composers have a sharp eye for interpreting the landscapes around them, just as, theoretically at least, geographers have an equally sharp eye for interpreting those same landscapes—though not necessarily music! The main difference, of course, is that cultural geographers' "compositions" are expressed in words, not musical notations.

3

Waterscapes: Toward the "Sea of Tonality"

"A MOUNTAIN STREAM IS A CHORD OF MANY NOTES"

This chapter continues the theme of time and space from the standpoint of water. Water sculpts landscapes. Most water runoff is taken to the sea, whence it evaporates and is carried over the land to fall as rain or snow, whereupon it flows into streams and rivers and once more begins its return flow across landscapes to the sea. This hydrological cycle is repeated endlessly. As will be seen, composers have considered the movement of water at varying stages of the cycle. Moving water exhibits and reflects the strength and power of nature, a fact composers have not missed. Nor have they missed the point that rivers, lakes, and seas also reflect intangible qualities attributed to them by human societies. For the purpose of this chapter, the resultant wonderful soundscapes are best referred to as waterscapes.

R. Murray Schafer's (1933–) *Miniwanka—the Moments of Water* (1971), a choral work, uses North American indigenous people's words for "water" to trace its path from tiny raindrops, through streams, rivers, and lakes, finally, to its arrival at the sea. And in *Carved by the Sea* (2002) by another Canadian, Robert Evans (1933–2005), countless words for the sea are sung, backed by percussive sounds and the recorded ocean pounding the shores of Prince Edward Island. Using words is one thing; using just music is another, for evoking an image or expressing understanding of a scene or a physical process is a greater challenge to a composer who uses just the instruments of the orchestra.

A good place to start one's appreciation of a waterscape is in the countryside. Whether near the start of a river or further downstream, many nuances are revealed as one listens and observes the variety of soundscapes created by gently or madly rushing water. To the observant, it is wondrous to listen to the huge array of sounds created by falling rain or by a flowing stream as it gets larger and becomes a river, before it flows into some other body, such as a lake or the sea. Even to hear just the start of the flowing process is special because, as Schafer (1977, p. 18) has put it, "A mountain stream is a chord of many notes strung out stereophonically across the path of the attentive listener." Many composers have sought to capture this "chord of many notes" and to represent the colors and moods of water as it moves from humble beginnings via streams and rivers, and perhaps lakes, toward the sea, eventually mingling with water already there, sometimes by itself, at other times with humans and other creatures present.

COLORS LIKE A GENTLE BREATH OF WIND

Tōru Takemitsu often sought to capture flowing water in his music. *Rain Coming* (1982) opens with a simple figure played mainly on a flute. The tune metamorphoses several times in the orchestra in an effort to reflect the way water changes as it circulates in the universe. In another work, *Riverrun* (1984), for piano and orchestra, Takemitsu creates "a musical tributary" flowing away from the main current, "wending its way through the scenery of night toward the sea of tonality" (Takemitsu 1991, p. 6). These evocative words capture the essence of the superb rendering in music of the sound of flowing water. Another of his works, *Water-ways* (1977), reveals streams going off in their own directions, following different waterways, though in the end they come together and, as one huge river, flow "toward the sea of tonality." *Umi e (Toward the Sea)* (1981), for flute, harp, and strings, also by Takemitsu, is loosely reflective of Herman Melville's writings inasmuch as its three section titles are "The Night" (a serene, imaginary landscape), "Moby Dick" (a vision of a man of character), and "Cape Cod" (a sublime image of dawn over the tranquil sea). Melville's work inspired Takemitsu, but Takemitsu's imagination and skill created the sensitive and, yet, evocative orchestral soundscapes, which he described as fragments thrown together, as if in a dream. Some fragments! Some dream!

Takemitsu's appreciation for both stillness and movement were profound. Not only was he clearly touched by his observations of nature, but he took delight in conveying to others, via players' performances of his music, the sense of wonder and peace he gained from his surroundings. Silence is important in Takemitsu's work, as is the sense of gentle movement. His disarmingly simple musical landscapes shimmer in their detail. The subtle tone-color changes captivate the ear and mind, for they are like unexpected, sudden, gentle breaths of wind on an otherwise breathless, hot summer's day. The breath of air tickles the hair on your arm for a second and then flows on. Takemitsu's tone colors change subtly, striking our imagination for a second, and then are gone, but the images of his waterscapes (and also his landscapes) remain in our memory—and we are enriched.

Takemitsu drew inspiration from close contact with most of the landscapes around him, whether a tiny garden in an urban area or a large stream in a rural area. He reminds me of a Japanese professor of geography with whom I once traveled in New Zealand. I had the privilege of sitting beside him as the bus drove through North Island. That seating assignment proved fortuitous, for I was thus able to observe how he looked at the landscape. Rather than looking at the panorama and saying, "Wow, what an amazing view," as did most others on the bus, he looked at small details, such as a gate and fence on a farm, the steps to a house, a flower bed in a front garden, or the roof lines of neighboring houses, and whenever we stopped, even if for only a minute until a traffic light changed, he would quickly sketch the scene of the moment. Each sketch was wonderful for what it captured. I realized that I had been guilty of looking at the panorama and ignoring the details; hence, I began to focus, to be concerned with details, and to marry the small with the large. I am confident that Professor Shinzo Kiuchi ended the trip with a greater sense of the changing, intimate character of the countryside than did the others present. My new focus on the detail, even while considering the grand sweep, led me to a new appreciation for both landscapes and, later, music. Takemitsu's, and perhaps also Professor Kiuchi's, many small details in generally still and tranquil situations, when combined, create a sense of wholeness.

Concern for detail surely inspired Canadian Hugh LeCaine's (1914–1977) *Dripsody* (1955), an extended composition (on one of the very early tape recorders) based on the theme of the sound made by one drop of water dripping, repeatedly. It was a technological landmark and stands as an early example of *musique concrète*. Asked why he had titled the work as he did, LeCaine replied, "Because it was written by a drip" (Hugh LeCaine, Canadian Music Centre 2004). Most composers have been concerned with a fuller sweep of water flowing, with chords of many notes created by streams and rivers, including Canadian Gilles Tremblay in his picturesque *Fleuves* (1976). It is at the scale of rivers that we will continue.

COLD, CLEAR WATER

Finland's most famous composer, Jean Sibelius (1865–1957), was a genius for his representation of water flowing. As with Takemitsu, Sibelius experienced delight when in the countryside. His works reveal his understanding of the power of nature in Finland. In contrast to Takemitsu with his concern for details, Sibelius was more concerned with the broader characteristics of water in the landscape, especially with the pulsing rapid flow of water. Sibelius created a sparkling landscape in Symphony no. 6 (1923). It is not just Finns or other Scandinavians who appreciate the musical representation of "pure, clear water"—to use the composer's own words—that flows over ice-cut landscapes; listeners in Canada, Alaska, alpine Chile, and New Zealand's South Island can hear their own glaciated, riverine landscapes resonating in Sibelius's music. As far as we know, his music does not describe any particular ice melt and associated rivers, but the images created are so crystal clear that we need not know if they represent this or that specific place in Finland. It is enough to enjoy the sense of flow, of movement across the ice-carved landscape, of the bubbling, charging water. As in much of his work, the strings are dominant throughout, carrying the musical argument. Unlike many of his other symphonies, Sibelius excludes massive climaxes, choosing instead to focus on glittering orchestral sonorities. There is a strong pulse in the scherzo, which Robert Layton (1983, p. 55) calls a tour de force. Guy Rickards (1997, p. 160) feels that the work captures in sound "the rarified, limpid quality of the light of the Nordic countries . . . [with] a luminosity in its textures that . . . depict[s] the streams and forests of his country in the full flood of spring."

Sibelius struggled as he wrote this symphony, Symphony no. 6. It is not clear why he pushed himself to complete it. Perhaps his wife said she would leave him unless he finished the work, being fed up with his drink-induced rages (Rickards 1997). We must balance this sad comment with the acknowledgment that Sibelius was extremely sensitive to the natural world around him. Indeed, to share nature with him was a wonderful experience (Hill 1947, pp. 10–11). As he was completing the work, in 1917, Finland was plunged into a civil war. Despite the turbulent times, Sibelius stated that Symphony no. 6 was "very tranquil in character and outlook," that it captured "the 'rare sense of rapture' he had experienced as a boy" in a summer rural setting, and that "there is a pronounced mystical atmosphere" (quoted in Rickards 1997, p. 160).

Sibelius was not the only Finnish composer drawn to water in his compositions. Selim Palmgren (1878–1951), seen by Sibelius as a rival, never achieved the quality of composition achieved by Sibelius. Palmgren's most famous representation of a river was in his Symphony no. 2, *The River* (1913). Neither Sibelius nor Palmgren seem to have been representing particular rivers. Rather, they focused on rivers in a vast, glaciated landscape. In contrast, if we move to central Europe, we will find a river represented in music that is not only quite specific in its landscape depiction but also is surely the most famous river in music, the Vltava.

VLTAVA: NATIONAL IDENTIFICATION

The Vltava (to use the river's Czech name) or the Moldau (its German name) is part of Bedřich Smetana's (1824–1884) *Má Vlast* (discussed in chapter 5). The second and most famous in the cycle of six symphonic poems that comprise *Má Vlast* (1872–1879), *Vltava* is one of the most geographically descriptive of all compositions. The river is powerfully ingrained in the national psyche of the Czech nation. In that sense, it is more important than any other river in the Czech Republic. As portrayed in *Vltava*, the river starts as a vivacious, two-part, youthful figure before it assumes growth and increased stature as it flows through the countryside, passes important places, and then flows out of Czech territory and, so, beyond the musical picture. Smetana outlined the contents of the music in a letter, which is partially paraphrased here, with some additional comments.

The Vltava river has its origins in two small streams in the Sumava forest in southern Bohemia: two flutes and then clarinets represent the streams. One is cool and calm, the other is warm and vivacious. The listener experiences the sparkle and ceaseless motion of the water, so vivid is the playing by the woodwinds, joined by strings and harp. The music reminds Czechs that, as a people, their beginnings are deeply rooted in the past. Those beginnings were once young, but as time progressed, their sense of being, their sense of national self, grew and needed nourishment, just like a growing river. The streams weave together and flow on as a young river, symbolized by the strings, through the lovely Czech countryside, filled with groves and meadows. A hunter's horn sounds, and the river flows past a happy wedding feast and the joyous strains of a polka. Then, beneath castle ruins proudly perched on nearby cliffs—which remind people of their heroic past—the river's strong but steady flow is revealed, a reminder of the core strength of Czech nationalism. However, the river's path is soon broken by the St. John Rapids (today hidden due to flooding following the construction of a hydro dam downstream). The turbulent waters are vividly depicted in a dramatic interplay involving all sections of the orchestra, surely to present a reminder to Czechs that things have not always been easy for them. That musical outburst soon subsides and is followed by a powerful, full-orchestra restatement of the main theme. There is a sense of majesty as the river reaches and passes Praha, past the high castle of Vyšehrad, and, as a large river, *Vltava* flows off beyond Czech territory and, so, into the distance to join the Elbe, thence north to the North Sea.

The readily appreciated series of landscape pictures Smetana creates blends great sensitivity (especially at the birth of the two streams and during the night's moonlit flow past the field where nymphs dance), joy (the celebratory wedding dance at the river's side), and, especially, national pride (the powerful river as it passes Vyšehrad). The ending of Smetana's work contrasts markedly with the delicate manner in which he constructs the twin origins of the Vltava. Two gigantic chords played by the full orchestra to end the work refer to the opening image of the two streams struggling to flow over the rocks and down the slope to join and, so, continue their adventure together. The work embraces a delightful irony, for while Smetana made use of a variety of Czech songs in this and the other five symphonic poems, the main theme used in *Vltava* derives from a Swedish song, not a Czech folk song. This is of no great consequence since most European nationalisms were created in the nineteenth century during a time when there was considerable cross-fertilization of ideas across the continent. *Vltava* (usually referred to as *The Moldau* in North American concert programs) is often performed as a separate work (i.e., without the other five tone poems), but it is unquestionably best heard as part of the complete cycle that makes up *Má Vlast*.

A MAGICAL LAKE

The arrival at a lake can lead to interesting insights. An early twentieth-century work by American John Alden Carpenter (1876–1951), *Adventures in a Perambulator* (1914), conveys a child's being pushed around by the nanny one afternoon. Along with "The Policeman and Dogs" and other findings, the music delicately reveals the discovery of "The Lake," with the flute representing little waves and the strings and the horns sounding out larger waves. This work is concerned with but part of the daily cycle of time. A fuller sense of the daily cycle as it involves the movement of the sun finds expression in another work, one by a British composer who visited another lake.

While on a trip through parts of Africa, Michael Tippett (1905–1998) was powerfully struck by the way the sun at noon transformed a small lake's whitish-green color to whitish pink and then, as the sun descended, back to whitish green. He thereafter wrote *The Rose Lake* (first performed in 1995), which sensitively expresses in music the remarkable visual transformations he witnessed. Following a gentle opening section, the lake begins to sing (horns) as the sun rises; the lake's song is echoed from the sky. The song (played by full strings, with other instruments joining in during various blocks of sound) grows and sings throughout the day. In time, the song must leave the sky as the close of day approaches. The work ends with the lake singing itself to sleep. To help achieve a sense of the mystery and wonder of the place, Tippett's orchestra includes two harps, six horns, and a large array of percussion. The percussionists seldom play loudly. The most demanding task in the work is given to the percussionist who must play numerous roto-toms (small tunable drums) covering three octaves. The instruments are arranged in a semicircular pattern and have to be retuned several times as the work proceeds. It is extremely difficult to play roto-toms with the degree of subtly demanded by Tippett's score, but the effect when it is performed by an excellent player is stunning, for thereby the ingenious scoring succeeds in conveying a sense of the color changes Tippett saw that day he was at the lake. A listener will need the written word to appreciate Tippett's source of inspiration for the work because, though it is wonderfully expressive of a transformation in nature, it does not obviously convey images of a lake as such. Once the story behind the composition is known, however, the music "reveals" the many changing images for the listener. The changing sunlight on the lake inspired Tippett, but, of course, it was his sensitivity and skill as a composer that gave birth to a remarkable musical picture. Interestingly, *The Rose Lake* was cocommissioned by the London, Boston, and Toronto symphony orchestras. Each orchestra "premiered" the work in turn.

THE BEAUTY AND POWER OF THE SEA

The sea forms a significant part of the hydrological cycle. Numerous composers have sought to capture the ever-changing moods and remarkable energy of seascapes as wild nature and as a setting for humans against nature. All who have been on the high seas know to be respectful of nature and attentive to the potential for storms. Whether based on experience or imagination, many compositions communicate the various moods of the sea: the ebb and flow of tides, a gently undulating surface with nary a breeze, sprightly waves with fountains of spray, huge waves with whitecaps surging ahead as winds whip them along, the mighty roar of water crashing onto a shore, or, in contrast, the gentle lapping of dissipated waves as they caress rocks or flow onto a sandy beach. On the other hand, I am not aware of any composer who has been able to portray the buildup of a sand dune, so, perhaps, there is a limit to landscape as inspiration.

Anyone who stands on the seashore or on the deck of a ship is aware of the constant in and outflow of waves, with the character of waves changing significantly as weather changes occur. The sounds heard can sometimes be soothing, at other times very threatening. How have composers captured the sounds and moods of the sea? An essential character of most such music is of increasing and then decreasing swell. This is conveyed by both a sense of rhythm and increasing and decreasing volume. Most composers convey a feeling of regularity in the changing swell, which is fine for many a sea. However, waves do not have an absolutely common tempo to them, as R. Murray Schafer discovered. In his String Quartet no. 2 (*Waves*), Schafer creates dynamic, undulating wave patterns, the rhythm and structure of which are based on his analysis of wave patterns off both the Pacific and Atlantic coasts of Canada. Incorporated into his music is his discovery that the duration from crest to crest is between six and eleven seconds. To ensure that performers get his meaning, Schafer has written into the musicians' scores the varying wavelengths. As a result, when played, a fascinating musical seascape emerges that reflects an appreciation for the irregular regularity of the sea. Most orchestral composers seem unaware of such variations, for their music is generally consistent in time, without encompassing the wavelength variations discovered by Schafer.

MOVEMENT IN THE SEA BY THE MASTER IMPRESSIONIST

Not one to be taken up with the mechanics of time, Claude Debussy (1862–1918) painted several beautiful impressions, or images, of the sea. Chapter 4 discusses the first two of the three parts of *Nocturnes*; at this point, we consider the final part, called "Sirènes" (sirens being mythical half-woman and half-bird creatures living on rocky isles, playing on moonlit waters, and luring unwary sailors with their enchanting singing). A mermaid chorus joins the orchestra: no words are used, just a single vowel, rising and falling with the orchestra (mostly strings, woodwinds, and harp)—or, rather, rising and falling with the sea—until the chorus and two harps hold, while a horn solo reminds us briefly of the first nocturne, "Nuages," thus giving the three-part work both unity and closure; then, all trails off into a mysterious silence. When the piece is well performed, the effect is ethereal. This amazing work explores finely nuanced sonorities of delicate, pastel tone colors.

Debussy's ability to capture the movement of the sea with miraculous precision, vivid imagery, and beautiful colors led to his reputation as the most effective impressionist of the sea in music. His skill was well expressed in a later work, *La mer: Three Symphonic Sketches* (1903–1905). In this work, Debussy chose to contrast a quiet sea with a turbulent one. He thereafter composed no further programmatic music as such, but he did continue to write music about "my old friend the sea," as he put it (Debussy, cited in O'Connell 1950, p. 127). *La mer* is a work of considerable spiritual dimension. As he wrote it during an especially difficult time in his life, we can wonder if the array of changing intensities in the music reflected his own emotions. The three symphonic sketches superbly, subtly, and magically capture the essence of motion in the sea. Jonathan Kramer (1988, p. 230) observes, "Just as sea waves move toward the shore while the actual water bobs up and down in place, so the melodies in 'La Mer' move forward in time while the underlying harmonies and sonorities are nearly static."

Debussy's first sketch, "From Dawn to Noon on the Sea," opens with the image of the ocean gently at rest in a moment of mystery at dawn. A wind comes up, and we hear a series of very quiet, delicate, rhythmic and harmonic fragments—first on basses and timpani, then joined by two harps, muted strings, bassoons, clarinet, cor anglais, oboe, and muted trumpet—as the sun slowly rises, as if through a mist. Using various instruments, with harp and strings rising and falling as other instruments create impressions of movement, it is wondrous to hear and consider the resulting mosaics of ever-changing sunlight shimmering on the water, clouds reflected in the rising sea, and waves tossed by the wind. The sense of initial remoteness goes as horns and timpani create swells of sound, with strings suggesting the wind impishly blowing here and there. Throbbing rhythms of the sea's movement are fascinating to listen to. At the end of the scene, the wind rises to a climax but then relaxes, and the music ends quietly. It is doubtful that anywhere else in music is the sea captured so beautifully. Indeed, Debussy's creation of a sparkling, moving sea is stunning in its beauty.

In the second sketch, "Play of Waves," the sea grows from a state of calm to one of playful animation, with wind-whipped orchestral waves and brightly tossed spume colliding and sparkling in merriment, before fading, as if weary of the sport, to stillness. Again, the harp is important for its character. It is not a solo instrument. Rather, it is almost miraculously woven into the total picture. Debussy's waves clearly are special.

In the final sketch, "Dialogue of the Wind and the Sea," a storm approaches, and soon, fierce winds whip the sea into wild fury. The sketch opens with timpani and bass drum rolling quietly, periodic tam-tam blows hinting at something more to come. Cellos, basses, and then violas join in, with the higher strings and wind instruments playing against each other to create the image of interplay between the wind and waves. We sense a storm is coming, but just when we expect it to break into full fury, a sense of relief descends as representations of sirens' voices are heard (played gently, in turn, by woodwinds, harp, and brass). The orchestra then combines all of its forces to drive to a dramatic conclusion.

The above comments on the programmatic character of the works are mine, not Debussy's. They are based on his suggestive titles and my reading of the full score. Listeners may hear other things, such as the cry of gulls, or an image of the sky as it influences the color of the sea, or maybe waves seething around rocks near the shore. Whatever words one uses, *La mer* is varyingly lovely and terrifying, peaceful and hostile, dazzling and dark, capricious and steady, familiar yet remote. Magically transparent musical colors dazzle the ear and the imagination. Debussy's gift was his ability to write music that both astonishes the intellect and creates images in the mind. Interestingly, critics add another dimension when reviewing recordings of this work by sometimes including comments as to whether a conductor has placed the music in the Mediterranean, the northern Atlantic, or the Southern Hemisphere (March, Greenfield, and Layton 1999, pp. 403–5).

Debussy knew *La mer* was innovative: "It may be in fact that I have taken on a task that is too much for me; there is no precedent, so I am obliged to invent new forms" (Debussy [n.d.] 1997, back cover). *La mer* certainly does not present a sentimental or a romantic picture. It is magnificently impressionistic: the richness and variations in colors convey the image of a wondrous, ever-changing sea as the winds change in intensity and the sun slips in and out of the clouds.

THE SEA TO TWO ENGLISHMEN

English composer Frank Bridge's *The Sea* (1911) is worth mentioning for two reasons. The first is simply that the four-part suite is a fine example of impressionistic musical painting. Bridge described the first section, "Seascape," as an impressionistic painting of the sea on a summer morning, with the swell gradually rising and falling. The great expanse of water lies in the sunlight, warm breezes playing over the surface. The music evokes a relaxing feeling as the composer tightens the tension and then releases it as the swell lessens. The section entitled "Sea-foam" follows with the bassoon and strings, and occasional horn and other woodwinds, creating for us the lively scene of froth bubbling among some low-lying rocks and pools at the shore. Then comes nightfall and "Moonlight." Bridge commented that moonbeams (woodwinds) struggle to pierce dark clouds (strings and horns), which eventually pass over, leaving the sea shimmering (flutes and harp) in full moonlight. The work ends with "Storm." Wind, rain, and tempestuous seas (with the crests of waves smashing ahead as brasses blare, strings and timpani crescendo, and cymbals crash). A lull in the storm is to be regarded, in Bridge's words, as "the sea-lover's dedication to the sea." The second reason for noting this work is that it caught the imagination of young Benjamin Britten (1913–1976).

Britten was thrilled when, as a ten-year-old, he heard Bridge's *The Sea* (J. Burke 1983, p. 97). The impression stayed with him, partly because Bridge later became his teacher, and Britten, in turn, became Bridge's champion as he sought to ensure the continuing performance of Bridge's music. Britten also knew the sea, especially around Norfolk and Suffolk on the east coast of England, and he wrote about it in his operas *Billy Budd* and *Peter Grimes*. The several parts of *Peter Grimes* are linked by interludes, which serve as transitions. Due to their popularity, Britten assembled four of them as a single orchestral work called *Four Sea Interludes* (1945). "Dawn" arrives (flutes and high strings in a haunting tune), with short intrusions by screeching birds (woodwinds, harp, and brass). "Sunday Morning" follows: a fishing village, with a depiction of the sun shining on the sea and the harbor. Church bells peal (the horns backed by staccato woodwinds) as gossiping villagers (conveyed by a lively interplay of all instruments) begin to go about their daily activities. All too soon, the hustle and bustle subside, and in "Moonlight," a sense of brooding is conveyed as the moon shines on the water (bassoons, horns, and lower strings are dominant). The work ends with "Storm" (a thunderous interplay among the brass, timpani, and the remainder of the full orchestra). The sense of agitation is palpable, representing as it does, in the opera, the locals' sense of powerlessness after the hero is lost at sea.

SEARCH FOR NEW MEANING

Whereas some composers are descriptive of the sea or tell stories about people in relation to the sea, others seem more intent on using the image of the sea for what it reveals in terms of meanings. For example, Ralph Vaughan Williams's huge Symphony no. 6, the *Sea Symphony* (1947), reveals something of the composer's personal search for meaning. This immensely powerful work is programmatic, revealed largely through the use of words for soprano, baritone, and chorus. References are made in sound and song to sea shanty tunes and various other works on the sea. So deep and troubling are the inner voices in this work that Hugh Ottaway (1972, p. 47) described it as "the humanist apocalypse" in which, by the end of the first movement, "the only human positives are endurance and acceptance."

English composer Frederick Delius used poetry as his reference for his pleasantly romantic *Sea Drift* (1904). The American composer Richard Rodgers's *Atlantic Convoys*, in contrast, was inspired as much by movie documentaries as real events (chapter 7). Set during World War II, this work boldly seeks to reveal his understanding of what sailors went through while on ships during dangerous times.

There are many other works that could be discussed, including Arnold Bax's *Tintagel* (1943). Bax wrote in the foreword to his score that the work is intended to evoke a tone picture of the castle-crowned cliff of Tintagel and, more particularly, the wide distances of the Atlantic as seen from the cliffs of Cornwall "on a sunny but not windless summer day." Into this scene, he imposes images of mythic characters to create a vibrant sense of movement.

The Japanese composer Toshio Hosokawa's (1955–) *Memory of the Sea* (2000) deserves special mention. He grew up in Hiroshima. His Hiroshima was the recovered city after its destruction by the atomic bomb. As a youngster, he was powerfully struck by the quiet, forceful, invisible force of nature that brought forth new life following the horror and devastation of sudden destruction during World War II. *Memory of the Sea* is an ode to nature's power, for new trees, flowers, and wildlife, all of which reveal nature's remarkable ability to regenerate itself. Hosokawa's Hiroshima is thus a place of life, of pleasant images and warm memories, and, so, it stands in stark contrast to scarred memories of people a generation earlier whose world literally was blown up (see pp. 163–64). Hiroshima is on the sea: the sea persists; it was there before "the bomb," and it is there now. Hosokawa's *Memory of the Sea* focuses on the sea around the city. Using a contemporary style for this work, he augments the standard symphony orchestra with two groups of Japanese string instruments. The latter are located in the hall in such a way as to create a three-dimensional echo effect. The work expresses in music Hosokawa's childhood memories of sea breezes, moving clouds, and changing qualities of light. From *Memory of the Sea* and the other works just cited, we get a sense of the ebb and flow of the sea as time and space are mingled in ways unique to each composer.

PEACE OF MIND, WITH WORRY

Ludwig van Beethoven and Felix Mendelssohn wrote works on the very same topic. In English, the title translates as *Calm Sea and Prosperous Voyage*, based on two poems by Johann Wolfgang von Goethe. It is not clear if Mendelssohn knew of Beethoven's composition (though R. Larry Todd [1993, p. 59] maintains he must have had a copy of Beethoven's score in front of him as he wrote). Both were inspired by the same text, but while Beethoven (in 1815) felt confined to keeping Goethe's poem in his choral setting, Mendelssohn (in 1828) felt free to compose without the words. In so doing, Mendelssohn supported A. B. Marx's (1865) contention that instrumental music was inescapably becoming programmatic. The following poems were the stimuli for the two works:

Calm Sea
Deep silence broods over the waters;
the sea rests without movement;
and the mariner is troubled at the sight of the smooth levels around him.
No breath of air from any quarter;
the deathly silence is awful;
throughout the enormous distance not a ripple stirs.

Prosperous Voyage
The clouds are torn apart, the sky is clear,
and Aeolus takes off the chains of anxiety.
The winds rustle, the seaman bestirs himself;
hasten, hasten! The waves divide;
the distanced approaches;
already I see land.

There being no coast to see, anxiety prevails as the "deadly silence" and dead-calm sea stretch "throughout the enormous distance," but relief quickly takes over once land comes into view. Mendelssohn's orchestral rendition has greater color and character than does Beethoven's version. The latter uses the words to convey images whereas Mendelssohn's *Calm Sea* presents a remarkable, static image, which Todd (1993, pp. 58–59) states is due to tonal qualities, the use of pedal points—sustained notes held in one voice, usually the bass line (e.g., by the basses, cellos, or bassoons), while other voices move above it (e.g., violins)—and repeated phrasing rather than strong thematic development. The dead calm is broken by a flute solo, which is followed by a lovely impressionist soundscape as the boat starts to move and land comes into view. Leon Botstein (1991, p. 27) feels that Mendelssohn's "visual imagery" is able "to depict the sense of distance, color, light, and three-dimensional space through orchestration." Further, Botstein (1991, p. 27) seems to endorse the static nature described by Todd, inasmuch as he maintains that the "sections of the work follow as would a series of sequential paintings, which do more than chronicle action." Mendelssohn knew the sea quite well, for he traveled by sea from Europe to Britain, en route to Scotland, where once more he went on to the sea. He also traveled well, generally being hosted by royalty and landed gentry wherever he went. One of his best-known works resulted from his travels.

A SCOTTISH CAVE

It is not always clear that composers are describing a specific shore or experience; however, we know that some composers write works reflective of specific, real scenes. The best known is by Mendelssohn, who, while on a walking tour through Scotland in 1829, was thrilled by the grey sea surging, crashing, and hissing in and out of a huge cave on the weather-beaten, uninhabited, west-coast island of Staffa. There is a strange grandeur to the black, columnar rock formations (which resemble a grand pipe organ) out of which the cave has been carved by the sea over hundreds of years. Mendelssohn immediately thought of—and later wrote out, when he reached Glasgow—about twenty bars of music, which became the opening theme of the overture published in 1835 as *The Hebridean*, also known by its alternate (and popular) title, *Fingal's Cave Overture*. Mendelssohn was an excellent painter and sketcher, but he was too seasick while on the trip to Fingal's Cave either to draw what he saw or to write out his first thoughts, though his sketch of Oban (the departure port on Scotland's west coast) seemingly survives to this day. Marx implies that his ability to sketch music drew upon his marvelous sense of color when painting (Marx [1865] 1991, p. 212).

The opening theme beautifully captures in orchestral sound the ocean flowing rhythmically in and out of the cave. The theme, which appears many times throughout the work, always has this repeated rhythmic (crescendo-decrescendo) quality to it, creating the image of the ocean's surges against the small island and into and out of the huge cave. A second theme evokes the image of waves whipped into a fury, lashing the rocks. Finally, the storm subsides and works into a quiet recapitulation of the first theme, before the work ends with an animated coda. A geomorphologist likely could not describe the scene as vividly and effectively as does Mendelssohn.

Each summer, the large ferry boat *Columba* sails from Oban and goes past the island of Mull to Staffa, the location of the giant cave, and then on to the island of Iona. The latter small island has served as the ancient burial site for Norse kings and queens and an early Christian settlement; today, it is the site of a Church of Scotland abbey, where people go to study and reflect as part of their personal searches for meaning. While the ship heaves to just off Staffa's shore, passengers hear Mendelssohn's music over loudspeakers—a good blending of music, landscape, and tourism. Corny though it may seem to some, the music, coupled with the movement of the ship on the waves and wind blowing one's hair, adds to the sense of wonder one gets from seeing the grand cave as it is washed by the sea. Incidentally, the inner (usually return) route taken by the *Columba* partially mirrors that taken by Mendelssohn, whereas on the outgoing trip, on good days, the ship will take the outside—ocean—route around Mull to Staffa. Mendelssohn also wrote about the sea in *The Lovely Melusina* and, commented on above, *Calm Sea and Prosperous Voyage*.

A final question here: is the *Hebridean* overture simply a good example of a musical geography, or is it something more than that? My first reaction was that it is a superb example of musical geography, given that one so easily appreciates the noise of water going into and out of the cave. However, as I reflected on the question longer, I came to accept that it deserves the wider term *geography in music* applied to it. Why? A first response might be that Mendelssohn completed field observations prior to writing his work, just as would a geographer. A second point is that not only does Mendelssohn successfully convey an appreciation for the movement of water, but he implicitly describes erosional processes (as the water washes into and out of the giant cave), so I changed my mind. The work is deceiving in its simplicity. The more one listens to the work—or reads the full score—the more one realizes Mendelssohn's genius in having composed a work that so cleverly captures the sense of movement of the sea and of erosion. Seeing the impressive physical feature sitting off Scotland's coastline, while matching the music to the actual movement of the sea as it flowed against or crashed on the rocks, confirmed for me that this is a supreme example of orchestral music that is explicitly geographic in character.

THE SEA AND STORMS

Some of the works discussed above mention storms. Likely, many composers are attracted to the sea by storms because of their challenge. A common theme seems to be to create a gentle sea, have a storm, enjoy a peaceful resolution.

The Finnish composer Jean Sibelius drew inspiration from the often rough voyages he took to see the wider Europe and when he crossed the Atlantic. Recalling these experiences, he wrote *The Oceanides* (1914), a stunning work that provides a gut-wrenching, grim, musical picture of churning northern seas. His biographer, Guy Rickards (1997, p. 118), maintains that *The Oceanides* is an extraordinary score, being "the subtlest, most magnificent evocation of the sea ever penned." Clearly, Rickards is biased, for there are many other contenders for such an accolade. Keep reading!

Sibelius was not alone in drawing from experiences while at sea. Hector Berlioz (1803–1869), who recalled a most unpleasant Mediterranean sea crossing in *Le Corsair Overture* (1844), provides another example. Berlioz was in a smallish vessel when a sudden storm blew up. The craft nearly capsized, and Berlioz and his companions were lucky to escape with their lives. The music is vividly expressive. The work opens with the sea breezes and sea foam flying through the air as the boat rises and falls with the waves. The score is telling, for it looks a bit like a mountain profile, with "waves" of music going up and down scales. The tempo changes, faster and slower and back, as the work proceeds. We may wish we had taken seasickness pills, so strong is the musical image of the surging waves and, at times, the violently rolling sea, as well as our fear that the boat will go down. Donald Tovey tells the story about the English landscape painter Joseph Turner, who had himself tied to a mast of a boat so that he could draw waves smashing onto the deck. Turner's resulting paintings (in London) vividly catch the mood of the violent sea. So, clearly, both artists, Berlioz and Turner, were inspired by living through frightening experiences. We are the beneficiaries.

The best-known storm at sea in music may be Richard Wagner's overture to *The Flying Dutchman* (1841), a popular orchestral piece that reveals a seascape in a turbulent and tragic mood. Drawing on his own experiences as a sometime sailor, and based on an age-old story of sailors' superstition and fear, the work considers an unfortunate mariner condemned to sail the seas endlessly in his ghostly ship. Stormy winds (conveyed principally by wild chords on the strings) open the work. The Dutchman's motif or leitmotiv (a recognizable melodic fragment that appears frequently in the work) is proclaimed by the horns and bassoon. There follows a highly dramatic storm, with screaming winds tearing at the sails, waves rising and then smashing on the ship, and fragments of the sailor's cry of fear amidst the turmoil. The gale subsides and is followed by calm passages: an air from a ballad (in the opera), a joyful sailor's dance, and the restatement several times of the ballad, symbolizing redemption by love.

Wagner's music painting is utterly different from that by Debussy. Whereas Debussy is subtle, concerned with tiny details as part of the whole, Wagner is bold and brash, concerned with the grand sweep. Yet, each created readily understood seascapes in sound. Such is the marvel of distinctive voices—autographs. One of the many composers who admired Wagner's genius, Russian Alexander Glazunov (1865–1936) wrote *Morye* (*The Sea*) and dedicated it to Wagner. His work opens with the representation of winds becoming strong, the sea growing agitated, and a sense of darkness, which leads to the development of the furious and powerful roar of a huge storm, which in due course subsides, to be replaced by the sun shining on the calmed sea. The work was strongly influenced by Wagner's sounds and techniques. Similarly, but of more recent origin, Danish composer Rued Langgaard (1893–1952) wrote Symphony no. 15, *Søstormen* (*The Sea Storm*) (1937).

Russian composer Anton Rubenstein's (1929–1894) richly textured *Ocean Symphony* (1851/1864/1880) reflects the composer's fascination with the sea and draws upon his several years of sailing while away from Russia. Many regard this symphony, a solemn work from which one gets a sense of profound loneliness, as Rubenstein's principal work. In it, the qualities of each of the seven seas are revealed. However, selected themes and rhythms, including what appears to be exotic bird calls—not sea bird calls—raise doubts about the seven seas notion in the ear of a discerning listener. Rubenstein originally wrote four parts but later added three more, including the one about a storm.

WITHIN THE SEA

As noted earlier, Debussy's *La mer* captures the sense of waves moving toward the shore, even "while the actual water bobs up and down in place." The attentive listener may be able to imagine what it feels like to be within a wave. This thought leads to the question of what it may feel like to be within the ocean. The Canadian composer Robert Evans wanted to find out by focusing on whales.

Evans (personal communication, May 4, 2000) began his composition process by immersing himself in recorded whale songs drawn from pods in several oceans and at different times of the year. His sense of wonder and awe increased as he listened to their many sounds. He also read all he could find about whales and about how humans are harming and even destroying certain whale environments with pollution. Only when he was satisfied with his knowledge base did he start to write *Whale Song Dancing* (1993), a work for recorded whale songs, four part choir, and chamber orchestra. The orchestra includes organ and synthesizer, the latter to create "some extra-human sounds." Evans calls it "a celebration of the sea, of whales of so many kinds, their song and dance and something of our relation with them." And then, of note, he adds something that many composers must feel: "It is personal, but it isn't private." According to Evans, the listener to this work is caught between opposites, "power and peace, nurture and death, timelessness and urgency, man and his ambivalent love/hate fascination for whales, these largest of the earth's tenants who move so elegantly and effortlessly in the oceans of our world." Evans's uses compline (the plainsong sung at the last of the canonical hours at Vespers) in different disguises as the work progresses. "It is intended as a metaphor for our frequent serious abuse and lack of concern for our environment, in particular the sea and whales." One section concerns humans lost at sea, composed in the form of a sea-shanty. Evans incorporates the whales' songs into the music, not letting them stand wholly apart. Ultimately, following a hymn of praise and wonder, the sea has the last word, as the work ends on a note of hope.

In keeping with the theme of this chapter, there is a strong sense of movement within *Whale Song Dancing*, as we hear the whales moving (including the recorded sound of an orca accelerating from a standing start to 35 mph, which, as Evans puts it, is "a watery Indy 500"). The words, by Evans, are written from the perspective of the sea and the whale.

The sea reveals

> I am the sea
> And all that heaves and rolls endlessly
> In the running of the swell
> Of fathoms below
> Where peaceful tends my watery garden
> Of teasing underwater whisperings,
> I am rolling sea
> And so my legends
> Stream and spill around
> The corners of my oceans.

And the whale reveals

> I am whale
> I move in rhythm with the sea,
> in whale time;
> Slowly

Evans does for us with whales within the sea what Franz Peter Schubert does with respect to a fish in a stream. Schubert's *The Trout* captures movement within the flow of a stream. It is for piano quintet, not an orchestra, but it should be listened to in terms of a waterscape for *The Trout* conveys a sense of stillness in the water and of the gentle motions of the fish in a truly exquisite manner.

Evans is not alone in being drawn to the wonder of whales. John Tavener (1944–) of England wrote a dramatic cantata entitled *The Whale* (1966), and American composer Alan Hovhaness (1911–2000) wrote *And God Created Great Whales* (1970). The latter work has a feel quite different from that created by Evans, yet both use recorded whale sounds. Of course, each composer has his distinctive style, with a unique autograph. Neil Stannard (1995, p. 2) calls the Hovhaness work "less 'composed,' even at times aleatoric." Sounding at times a bit too much like so many of Hovhaness's other works, the images created are nevertheless fascinating. Low strings growl and, with other instruments in the orchestra joining them, swell grandly, signifying the presence of underground mountains (though a listener might instead imagine whales swimming around and surfacing). A quiet, meditative section follows by way of introducing the amazing recorded voices of four great humpback whales. They are wondrous. The orchestra then joins and overwhelms the whales with some interesting play by the horns and trombones to imitate whales singing. The real whale voices return, different this time, accompanied quietly by the orchestra, the strings representing waves. The interplay is repeated, but then a loud beat of the bass drum leads to a new set of whale songs. A sense of distress prevails; at least, that is what comes to mind as I listen to the whales sing. I wonder, was the recording boat too close? But the tension is released, followed by some lovely interplay between the several voices, backed by very quiet strings. They are replaced by a peaceful "Oriental" tune played by strings and woodwinds that is overlain and finally overthrown by a jagged tune played by the brass. A powerful nervous energy develops as the orchestra members are instructed in their scores to start playing quietly in repetitive patterns, "rapidly and not together in free non-rhythm chaos," and to build to a "very wild and powerful" conclusion (Stannard 1999, p. 3). Stannard includes a quotation from Hovhaness, which suggests that I could have discussed this work in chapter 2, before the discussion of Franz Joseph Haydn's *The Creation*: of his own work Hovhaness states, "Man does not exist, has not yet been born in the solemn oneness of Nature" (Stannard 1999, p. 3). The work is a mix of nature, represented by the whales singing—or are they speaking to each other?—and Hovhaness's distinctive, but culturally confused, style of music.

Whales, of course, are not the only creatures in the sea. New Zealander Anthony Ritchie's (1960–) chamber orchestra work entitled *Underwater Music* (1993) is a colorful work in three movements that depicts seahorses, stingrays, and dolphins. Another New Zealand composer, Gareth Farr (see Shieff 2002, pp. 191–204) wrote *From the Depths Sound the Great Sea Gongs* (1996) for Balinese gamelon and symphony orchestra. The work opens in the fashion of Maurice Ravel, with "little eddies whipping up the water" before the mysterious core in which one hears, in Farr's words, "tiny echoes of things" (quoted in Shieff 2002, p. 201). Finally, many creatures colorfully inhabit Vaughan Williams' *Sinfonia Antartica* (see pp. 138–41), including killer whales, seals, and various birds, including penguins. A delightfully humorous highpoint in the symphony exists when Vaughan Williams refers to the penguins as they waddle across the ice.

GOD OF THE SEA?

King Neptune rules the sea. This we know from his "appearances" (rather akin to Santa Claus) whenever ships cross the equator. So, there is a king, but what about a god of the sea? Undoubtedly, most sailors will say there is a God, notably those on board ships caught in the middle of major storms, who find themselves praying even if they never otherwise pray.

Indeed, sailors are known to be sensitive to folklore about the sea and have long been given to entreating and thanking a "higher being" for their safety. And those ashore who care for those at sea are given to singing the famous hymn—adopted by many navies as their hymn—which entreats,

> Eternal Father, strong to save,
> Whose arm doth bind the restless wave,
> Who bidd'st the mighty ocean deep
> Its own appointed limits keep;
> Oh hear us when we cry to thee
> For those in peril on the sea.

Before providing "proof" of the existence of a god of the sea, I must first note that I have experienced the stress of being on board a small ship sailing in the midst of a damaging hurricane in the Caribbean and on another ship of similar size that barely missed being hit by a viciously twirling, huge water spout in the Indian Ocean. Tension and great concern were evident among all aboard these ships, but especially the first, for the captain had to sail us into the eye, so we knew we also had to sail out of it. The screaming of the wind as we sailed into the relative calm of the eye of the hurricane and then away again is, to this day, fresh in my mind. Indeed, I find that the storm music by, especially, Wagner, Bridge, Britten, and Sibelius fascinating to listen to for their superb representations of the violence one can encounter within a major storm at sea. The composers' storms subside just as nature's violence subsides. As nature's storms ceased to be dangerous to the two ships I was on, I recall numerous people on board saying, as they looked out over the just-calming ocean, "Thank God." So, yes, as an impressionable youngster, I was convinced there was indeed a god of the sea!

4

Specific and Generalized Landscapes

Michael Tippett's delight in the subtleties of changing light at Rose Lake inspired him to write the composition already identified (in chapter 3). Composers have responded on various scales, from describing quite specific sites in the landscape, in the manner of Tippett, to more generalized representations of broad regions. Some landscape references are quite direct, such as with Ottorino Respighi's intimate story telling, inspired by specific features of Rome's landscape (see pp. 88–90). Sometimes a composer uses some other artist's interpretation as the source for inspiration, as with Anthony Ritchie's reference to paintings by Graham Sydney and to his own knowledge of the actual landscape in *Timeless Land* (2004), which represents the astonishingly variegated, generally arid landscape of Central Otago in South Island, New Zealand. Other composers elect to write in a generalized manner about a landscape and, in so doing, may call up impressions of the general character of the place, as did George Gershwin in *An American in Paris*. This chapter discusses examples of specific and generalized landscapes, first by considering sunrises and sunsets and some examples of regional landscapes before turning to some rural settings and then to cities and industrialized landscapes.

DAY'S BEGINNING AND END

Morning is a wondrous time of day, unless it is pouring rain and cold. On a clear day, the impact of the sun on the spirit can be profound. Numerous composers have sought to capture this time of awakening. Denise

Hulford's *Genesis* (1995) for a chamber orchestra was inspired by sunrise on her favorite New Zealand beach. She believes her title is appropriate given that genesis means "the beginning, origin." One imagines being on the beach with a breeze blowing one's hair, seeing the tips of the waves coming in, perhaps watching driftwood move as the waves rush in only to leave, and experiencing the increasingly evident landscape as the sun rises and the day begins. Another New Zealand sunrise occurs where the sun first appears each day. To celebrate the new millennium, a group of people climbed Mount Hikurangi on the country's North Island so that they could experience the new year first, but it was cloudy, and they could not see the sun! In contrast, Christopher Marshall's *Hikurangi Sunrise* (2000) takes place on a pleasant day, with seagulls calling as they soar and wheel in the breeze, with waves below catching the sun's first rays. Strings, harp, and marimba announce the first glimpse of the sun, soon followed by oboe and flutes, before there is a wonderful interplay between the orchestral birdcalls and color changes that suggests energy-filled movement. The work ends with several strong chords and a flourish, indicating that the sun is now up and the day can proceed.

Russian Modest Mussorgsky (1839–1881) created one of the most famous sunrises in music in *Dawn on the Moskva River* (1880). So named by the composer, it is the prelude to the opera *Khovantchina*. It is often played as a concert piece. The music opens with a feeling of melancholy; it is a place at rest, with hints at the dark history soon to be revealed. Dimly at first, and then more clearly, we sense the sunlight illuminating ever more features of the landscape. The central feature lies ahead of us: the Kremlin's walls and the colored domes of churches glint and then glow in the rays of the rising sun. Suddenly, a new depth is added to the sound as the great bells call believers to early mass. The music masterfully matures into a full-orchestral sound that unveils the interplay of nature and a cultural landscape filled with an amazing array of representations of power and authority, before a meditative section carries the work to a reflective, rather than triumphant, conclusion. As the music is performed, it is easy to imagine being at the river's edge, watching the sun rise and so bring to life the history-filled landscape. However, the focus is on light and landscape, not on the people within it.

In contrast, we can consider a George Gershwin (1898–1937) morning. A Gershwin morning? I admit that I am taking a liberty here, for this specific landscape is fictional, though when it is portrayed onstage, it is very real indeed. In some staged versions of the opera *Porgy and Bess* (1935), Act III, scene iii does not open with Gershwin's scored music, as published, but with a delightfully expressive sixty to ninety seconds of sounds made by the actors as they take us from a community asleep to one in vibrant interaction. The curtain goes up on a darkened, still-slumbering village.

Slowly at first, amid otherwise absolute stillness, there is a repeated beat by an actor pounding wood with a large mallet. This beat is gradually built upon, the rhythm becoming more complex as other actors join in, hitting hammers, sawing wood, and banging on pots and pans. The combined sounds get louder and then meld into a captivating rhythm that is given an added dimension as people wave and shout "Good mornin'" as they emerge from their houses. The sun quickly climbs to give light and new life to the world. As the sounds reach a climax, with a marvelous sense of joy and release, everyone onstage bursts forth in song, accompanied by the orchestra. The transition from a village still and in darkness to one filled with joy and sunlight is quite delightful.

Robert Schumann's Symphony no. 3, *Rhenish,* was written as a response to the warm welcome he and Clara Wieck had received in Düsseldorf in 1850 and his delight in the fresh Rhenish air. Though the symphony is not descriptive of any specific place, the scherzo was originally entitled "Morning on the Rhine," and the final movement seems to refer to moving from the somber setting of a darkened building, perhaps a medieval cathedral, into the bright sunshine and a lively scene outside.

A more contemporary consideration of morning is provided by Rued Langgaard, the Danish composer. His 1948 Symphony no. 14, *Morgenen,* opens with a fanfare, followed by a section on unnoticed morning stars. There follows "The Bells of the Marble Church" and "Weary There Rise to Life," then, finally, filled with compulsive energy, the unmistakable realization of a new day's light and beauty. In contrast to the energetic setting in *Porgy and Bess,* the joy-filled final movement of the *Rhenish Symphony,* and the at first slow and then lively realization of a new day in Langgaard's music, Richard Strauss provides a solemn sunrise in his *Alpine Symphony* (see pp. 132–34), a work that concludes with a sunset, and it is to a sunset that we now turn.

Day ends, of course, and the sun sets. This event also has been captured in music, and not only by Strauss. One such work, "Sunset near the Plantation" from Frederick Delius's *Florida Suite* (1887), captures sunset and the sense of joy people can feel after a hard day's work. The work opens with the quiet and sensitive interplay of floating orchestral colors (played on the strings and woodwinds, including the plaintive sweetness of the cor anglais). We sense the weariness of people as the sun descends. Then, after a cleanup and the evening meal, the people are alert again and enjoy a time of play and spontaneous dancing in the twilight. The time comes when they go to their houses to rest for the night, though some—maybe it is only the players and the audience listening to the work—stay to watch the sun's final descent. The work's opening motif returns, and we again hear the beautifully textured shimmering interplay between strings and woodwinds, until the sun is fully gone. A common feel to much of the music on sunrises and sunsets is a sense of calm, often in a pastoral setting.

BEETHOVEN'S PASTORAL MASTERPIECE

Symphony no. 6, *Pastoral* (1808), is so famous that most people likely think Ludwig van Beethoven was original in his exploration of a pastoral setting. In fact, many composers had previously worked with the pastoral theme, with birds and other creatures inhabiting their creations. For example, in 1771, Luigi Boccherini wrote a quintet with "Une scène champêtre, où le chant des oiseaux se marie au son villageoise" (with a lark, a quail, and a cuckoo), and both Johann Sebastian Bach in *Christmas Oratorio* and George Frideric Handel in *Messiah* included "pastoral symphonies" as interludes; and so I could continue (see Orrey 1975, pp. 20–53). While Beethoven obviously drew ideas from the past, deriving part of his story line from Boccherini and undoubtedly also from a work entitled *Le portrait musical de la nature* by the Rhenish composer Justin Heinrich Knecht, it is nevertheless clear that he brought pastoral representation to its peak, and so, his work has special meaning.

The image of landscape in the early nineteenth century was one of "neoclassical artificiality" in which artifacts (such as bogus Greek temples, broken columns, nymphs, and gods) were deemed of greater importance than the environment in which they were set (Lowenthal 1985). In contrast, Beethoven's nature clearly was not artificial; he told a friend he was "only happy in the midst of untouched nature." However, Robert Schauffler (1947, p. 261) tells us that the landscape enjoyed so much by Beethoven was actually "half-wild woods and fields," which raises the question, what is the nature of nature? Beethoven loved to wander through the countryside in the vicinity of Vienna, where he got endless enjoyment from "the broad meadows, rocky clefts, elm-covered woodland paths, and murmuring, rushing brooks" (Osborne 1995, p. 97). Richard Osborne (1995, p. 97) tells us, "Throughout his life, Beethoven's dedication to the countryside was absolute." When there, he felt he was *in* nature. In Symphony no. 6 (1808), we are able to "see" the landscape readily appreciated for itself, unadorned by artifacts and "unsophisticated by art, literature, mythology, or fashion" (Schauffler 1947, pp. 260–61). Not only did Beethoven's mind-set lead him to write music that forever changed music, but he also thereby challenged contemporary constructs about nature. In sum, he so disliked what intellectuals and "sophisticates" had done to "nature" that, in response, he remade the conception.

The work's opening moment, described by Beethoven as the "Awakening of cheerful feelings upon arriving in the country," is peaceful, with a graceful simplicity and delicacy that endears itself to our hearts and minds. There is a sense of serenity and subtlety. "Scene by the Brook" follows. The orchestral portrayal of the gently flowing, clear water is exquisite: flowing eighth-notes in the second violins are the key, though two muted cellos repeat the sound an octave lower, even as the first violins sigh lazily, as though the sun were both hot and relaxing. The higher the tone of the instruments, the shallower the water, and "the more water the deeper the tone" (Schauffler 1947, p. 263). With respect to the portrayal of the brook, Beethoven was in the Vienna woods with a friend when he said to the friend something like, "It was here that I composed the scene of the woods, and the birds composed with me." Sadly, we do not know exactly what site he was looking at; nor can we tell exactly which section of music resulted from the inspiration gained during that visit to the stream. Even so, as we listen to the music, we know that the place was a quiet spot due to the delightful suggestion of bubbling water that ripples through the music. It is sunny, and the air is fresh. Near the end of this altogether pleasant movement, with remarkable skill, Beethoven has us listen to a nightingale (flute), quail (oboe), and cuckoo (clarinet) singing together as if in a serenade to nature itself.

In the third movement, "A Jolly Gathering of Country Folk," we hear a not-very-good village band playing dance music, first represented by the strings and then answered by the flute, bassoon, and oboe. The people thereafter dance to another folk tune, a bit slower and to a different rhythm, before the movement ends with a return to the first dance. The image of a happy group is catching. Beethoven's band was based on one he listened to as he quaffed beer! The seven "wholly unsophisticated peasants" who formed the band "at the 'Sign of the Three Ravens,' a tavern near Mödling on the outskirts of Vienna," are remembered in the symphony, but more, they no doubt additionally influenced Beethoven with their expression of nature in their "unadulterated Austrian folk music" (Schauffler 1947, pp. 266–67). Interestingly, Beethoven sometimes wrote small pieces for those musicians to play. "Live and enjoy life" may have been their motto, and it is this that Beethoven conveys in his delightful tavern band's slightly off-tune and awkwardly in-time rendition in the Sixth Symphony. Suddenly, however, a warning from the skies sends the merrymakers rushing for cover, and ever so quickly, a storm comes in and shatters any sense of peace and joy. The storm (described in pp. 135–36) is amazing to behold due to its immense power. The work reaches a peak in tension and sound, and then release comes as the storm subsides and moves into the distance. As the fury subsides, a bird sings again, we hear a shepherd's song of thanksgiving, and the listener can sense sweet delight as the freshly washed landscape comes to life again.

The fourth movement, "Gladsome and Thankful Feelings after the Storm," is joyous as the sun-filled landscape is again delightfully at peace. Beethoven uses a simple, yet highly effective, hymn tune to reveal that his landscape once again is filled with light and freshness. This particular image of nature, Beethoven's nature, is now embedded in the minds of concertgoers the world over.

The pastoral focus of Beethoven's composition was partly based on his Christian beliefs. A famous hymn he knew includes the following verse:

> When thru' the woods and forest glades I wander
> And hear the birds sing sweetly in the trees;
> And I look down from the lofty mountain grandeur
> And hear the brook and feel the gentle breeze.

Further, the autographed copy of the finale of the *Pastoral Symphony* had on it, in German, "We give Thee thanks for Thy great glory" (Osborne 1995, p. 97). Beethoven clearly evokes the symbolism of that statement and the hymn's verse in his music. We can be glad for the *Pastoral Symphony*, for it stands as one of the most beautifully descriptive works on nature of all time.

All in all, when we listen to Beethoven's *Pastoral*, an idyllic landscape is balanced with the bleak harshness of nature, symbolized by the powerful storm. This opposition—idyllic nature/violent nature—proffers a severe tension. Beethoven, in acknowledging the overcoming of idyllic nature by the storm, may have been extending his acceptance that humans cannot control nature and they are directly influenced by nature, propositions widely held in nineteenth-century Europe.

Beethoven influenced others to write pastoral works, including Louis Spohr, who wrote *The Seasons*, and Joachim Raff, who composed Symphony no. 3, *Im Walde*. Then, there is Carl Nielsen's Symphony no. 3 (1911), which contains an exquisite pastoral slow movement that "irresistibly evokes the rolling countryside" (Fanning 1995, p. 358). Another lovely work, Ernest Bloch's *Pastoral* from his *Concerto Grosso for String Orchestra and Piano*, expresses tranquility, presumably in the Swiss alpine landscape, which, once he was in the United States, Bloch recalled with pleasure from the days of his youth. A quite specific landscape feature inspired *Lincolnshire Posy* (1937) by the Australian Percy Grainger, who described the work as a "bunch of musical wildflowers" (Gillies and Pear 2001, vol. 10, p. 270).

ROCKY ISOLATION

Three works about specific rugged places are by composers of Canadian and Danish origins. First, Claude Champagne's (1891–1965) *Symphonie gaspésienne* (1945) reveals his appreciation of the landscape of the Gaspé region of Québec. He sought to transpose the scenery into musical form by combining sounds that evoked the idea of fog, boats' sirens, bells, waves, and flocks of gulls, with a dynamism that suggested the steep slopes of the landscape by the sea. The second is *North Country Suite for Strings* (1948) by Harry Somers (1925–1999), another Canadian. By using sparse textures and lean melodic lines, Somers evokes the bleakness and loneliness one can experience in the northern Ontario landscape. The third work, by Langgaard, was originally named the *Kronborg-Kullen Symphony*, but that title was dropped, and the work is now known as Symphony no. 10 (*Yon Dwelling of Thunder*) (1945). The place reference is to the Kullen Peninsula in southern Sweden, where Langgaard spent his summers, delighting in the magnificent and varied rocky landscape forever being carved by wind and sea. Part of the work conveys a huge storm and has been called "The Flying Dutchman over Kullen."

Bleakness elsewhere has inspired other composers. For example, R. Murray Schafer's *North/White* (1973) identifies the dramatic confrontation between Canada's northern wilderness and technological society in the south. This orchestral work opens in a quiet manner and conveys a sense of space and of loneliness. Just as the audience relaxes to the gentle sound, however, the sense of well-being is shattered by an off-stage snowmobile starting up and revving loudly. Lest you think this too contrived, my wife and I experienced such noise intrusion one day while cross-country skiing on a clear, blue-sky day in the quiet winter countryside north of Ottawa, Canada's capital city (Knight 1991). We could see or hear no one. We stood motionless. All we could hear were gentle wind whispers amid the branches of a nearby tree. It was a beautiful moment. Then, suddenly, from perhaps a quarter of a mile away, there was the screaming roar of a snowmobile. Our delight in the quietness and solitude was shattered. Our soundscape mirrored the aural insult in *North/South*.

MUSICAL GEOGRAPHIES

As identified earlier, musical geographies offer distinctive clues that quickly identify specific locations or regions. For example, John Burke (1983, p. 139) notes that Malcolm Arnold referred in a rousing march to the boathouse of Padstow Lifeboat by incorporating the discord of a foghorn. Rather than focus on such detail, I turn instead to grander visions.

Felix Mendelssohn derived inspiration for several of his works from his travels in parts of Britain and Italy. I have already identified Mendelssohn's brilliance at composing a particular spectacular geographic scene in *The Hebridean* overture (chapter 3). Another work,

Symphony no. 3, *Scottish* (1842), uses "Scottish" references to infer Scottishness. While inspired to start composing by his visit to the ruin of the chapel at the Palace of Holyrood in Edinburgh in 1829 and his wandering in parts of southern and western Scotland, he was not immediately able to complete the work. Indeed, the work languished for many years, during which Mendelssohn went to Italy and thereafter wrote the bubbly *Italian Symphony*. Before considering the latter, let me first say more about the *Scottish Symphony*.

In 1831, he expressed concern about his difficulty in returning to "my misty Scottish mood" so that he could complete the work (Ellman 1995, p. 130). But complete it he eventually did, in 1841, and he dedicated it to Queen Victoria. The first of the four movements of his *Scottish Symphony* opens in a dour manner, perhaps reflective of either the Scottish character or the mist on the bleak moors and hills over which he had traveled to get to Oban prior to sailing to see Fingal's Cave. There is a strong sense of isolation, but then comes a stormlike rage that could either be the sea pounding the shore, perhaps near Oban, or else a storm over the land. Either way, the feeling evoked is dramatic. A short return to the opening dourness ends the movement. The second movement contains a colorful, dancelike melody reflective of Scottish dancing, with the distinctive "Scotch snap" rhythm and cutoff. One wonders if it represents a party scene Mendelssohn witnessed in Holyrood Palace or at one of the stately homes he stayed in while in Scotland, including that of Sir Walter Scott, whose novels he had read. The third movement continues to use the distinctive rhythm and cutoff, but at slow pace, at times loud, before, in the end, a quiet, deeply felt gravity returns, reflecting, perhaps, either Mendelssohn's visit to the room in which Mary Stuart, Queen of Scots, was imprisoned in Holyrood or the spot in the palace where the Queen's favorite musician, Rizzio, an Italian, was murdered. The final movement is turbulent and troubled. It may represent a battle. Indeed, Jonathan Kramer (1988, pp. 413–14) notes that Mendelssohn originally marked the finale *allegro guerriero* (fast and warlike) but changed it to *allegro vivacissimo* (fast with great spirit), which led him to wonder if Mendelssohn originally had wanted to evoke the image of Highland warriors at war. If one listens carefully to the finale, it is possible to imagine said warriors in full charge against the enemy—the English, of course (but here I react as a Scot). The battle won, the pace slows as the victors triumphantly sing a wonderful development of the symphony's opening idea.

An interesting question to ponder is, did Mendelssohn's Scottish references make the work sound convincingly Scottish? Robert Schumann heard the symphony and was told it was Mendelssohn's *Italian Symphony*. He liked it as such, praising it for its "gorgeous Italian imagery," so, clearly, in Schumann's mind at least, there was nothing so distinctively

Scottish in the work as to countermand his belief the work was about Italy. But since Mendelssohn said it was about Scotland, and the use of the Scotch snap and the occasional suggestion of bagpipes imply just that, many sentimental Scots in Scotland and in far-off lands cherish the work for its suggestiveness. Another work with Scottish references is Hector Berlioz's rambling *Rob Roy Macgregor Overture* (1831). Scotch snap references are made, and there is a lengthy reference to a Scottish-sounding tune before the work ends with a heroic theme. Reference to four Scottish tunes is provided in the Walter Scott–inspired *Scottish Fantasy* for violin and orchestra (1880) by the German composer Max Bruch (1838–1920). Bruch never visited Scotland but believed that the violin and harp epitomized Scottish folk music; hence their prominence in this delightful work.

Mendelssohn, Berlioz, and Bruch laid the groundwork upon which Scottish composers could build, with more specific references to Scotland's landscapes and its people. These include, for example, Hamish MacCunn's richly textured *Land of the Mountain and the Flood* overture and *Highland Memories*. Other works include Granville Bantock's *Hebridean Symphony*, Hugh Robertson's *All in the April Evening*, Muir Mathieson's *From the Grampians Suite*, and Cedric Thorpe Davie's *Royal Mile* (see below). Of note, Bantock took a walking tour of Scotland's Highlands as part of his preparation for writing the *Hebridean Symphony*, a work "of brooding mystery and impetuous drama" in which the music gradually grows in intensity from a scene covered in mist to a storm, a battle, a love-lament, and a song of victory before the mists descend once again (Anderson 2001, p. 2). Now let us return to Mendelssohn, but in another region.

Mendelssohn's Symphony no. 4, *Italian* (1833) is about a distinctive cultural landscape but, fascinatingly, makes few references to people. While in Italy, he traveled by coach, and though he liked what he saw out of the windows, he chose not to observe closely or talk with the inhabitants "out there," likely because he traveled and stayed with Germans. On the other hand, he was inspired by the ancient Roman ruins, the Vatican, Venice, and the many landscapes in between. His resulting delight is expressed in the work, for there are many lovely tunes. Part of his reaction to what he saw surely reflected the warm, sunlit landscape, which contrasted so markedly with Scotland's wet countryside. Mendelssohn loved Italy. However, only the final movement includes a distinctive regional flavor, expressed by the use of an Italian saltarello, a folk dance. This symphony, Mendelssohn's most performed work, represents a high point in nineteenth-century music, combining as it does classicism and romanticism.

Another great work, Mendelssohn's Symphony no. 5, *Reformation* (1830), includes a major quotation of the Protestant reformer Martin Luther's magnificent hymn, *Ein Feste Burg ist unser Gott* (*A Mighty Fortress Is Our God*), and makes use of the "Dresden Amen." The work itself is not directly about the landscape but of the Lutheran Reformation, achieved against the dictates of the Roman Catholic Church. The first movement reflects the reformers' struggle; the second is a pleasant scherzo; the third is a tuneful slow movement; the final movement is the effective victory call of reformation. It is in the final movement that the *Ein Feste Burg* theme has its clearest expression, first as a flute solo before being taken, in full sonata form, to an immense full-orchestral, magisterial statement of faith. There is a landscape tie, in hindsight. Anyone who has been in Geneva, Switzerland, and has visited the churches wherein reformers preached and seen the marvelous set of statues of the Reformation leaders in a beautiful garden setting will likely always thereafter think of that landscape when hearing this symphony. Though of Jewish heritage, Mendelssohn was raised a Lutheran. He was proud of both aspects of his spiritual identity, and he maintained ties to both. Interestingly, he drew upon Bach's work for inspiration, including the use of counterpoint (literally, "point against point," or the simultaneous setting of two or more melodic lines against each other, each line having merit on its own, yet with beauty springing forth when they are played together) in the opening and closing movements.

Another Reformation-related work, *Hussite* [*Husitská*] *Overture* (1883) by Antonín Dvořák, remembers John Huss and his followers, one of whom, in later years, was Dvořák. Interestingly, though Protestant, Dvořák also wrote *Stabat mater* (1877), involving the Stations of the Cross. It is to Dvořák that we now turn.

IN AND FROM THE NEW WORLD

Dvořák "wandered" from his homeland in Europe to North America, not as a tourist but as an invited guest. While away from his native land, he sought to write music different from that which he wrote while at home. His *From the New World*, Symphony no. 9 (1893), uses "Americanisms" to establish that he is not sharing thoughts about Bohemia. In 1885, a Mrs. Thurber in New York founded the National Conservatory of Music, and Dvořák was invited to be the director. Mrs. Thurber and he agreed that American composers needed to find a distinctive American—that is, U.S.—style of music, to be set apart clearly from anything Russian, or Czech, or French, or whatever else. He thus promoted existing American music and encouraged his students and others to search for their own distinctive voices. A colleague at the conservatory and one of his own students were instrumental in introducing him to Negro (the term used at the time for African American) spirituals and plantation songs. He recognized in them—and in Indian (Native American) songs—a potential source of distinctly "American" orchestral music. Dvořák did not copy any such works directly, but he was fascinated by the particular characteristics of this music, including its rhythms, harmony, and potential for orchestral color.

Despite what some people think, Dvořák's *New World Symphony* does not contain specific references to either Indian or Negro folk songs. Rather, he chose to incorporate his inner reworking of what he had learned. One of the tunes in his symphony is now so famous as an independent tune that many people believe Dvořák borrowed rather than wrote it. This tune, which became known as *Goin' Home*, was given words by one of his students in New York; hence, it can be said that Dvořák unwittingly "created" a Negro spiritual. Other points can be noted too. In the first movement, there are many folklike tunes, and there is a theme that sounds somewhat like the spiritual "Swing Low, Sweet Chariot." Dvořák acknowledged his use of Henry Wadsworth Longfellow's *The Song of Hiawatha* as inspiration for the inner two movements. However, he included themes in the third and final movements that are more reflective of Bohemia than of the United States, so even he could not fully escape his place and cultural roots of origin.

It must be observed that prior to Dvořák's pronouncement on the topic, American composer Moreau Gottschalk (1829–1869) was already doing what Dvořák recommended; he wrote huge, but now generally neglected, orchestral canvasses into which he incorporated Negro and Creole music (Chase 1966). In time, other American composers took up Dvořák's suggestions, in addition to jazz and other musical forms. Some were composers of consequence, namely, George Gershwin and Aaron Copeland, while some were lesser known, including Marc Blitzstein, Henry Gilbert, William Still, and Edward MacDowell. Other distinctly American music includes two rarely performed works, Leo Sowerby's *Prairie* and Virgil Thomson's *Suite for Orchestra: The Plow That Broke the Plains*.

How successful was Dvořák in conveying a sense of American identity in his music? Kramer is critical: "his understanding of American folk music was superficial. Our music was not in his blood" (1988, p. 258). Perhaps Kramer is correct, but Dvořák gave something special to American musicology by writing his symphony and by strongly encouraging Americans to draw upon what he regarded as a strong potential core for what could become distinctively American music. A second point worth noting is that, ironically, over the decades American audiences have loved the *New World Symphony*, even while choosing to ignore compositions by Americans. Finally, we must recall that the full title is *From the New World*, not *For the New World*. In other words, Dvořák wrote the work to take home, to give European audiences a feel for what he had gained while living in the United States.

FOLK MUSIC

Much of Dvořák's music has a distinct feeling associated with his native land, as in his *Slavonic Dances*, which, when heard, can evoke an image of a joyous rural gathering of farmers and their families as they celebrate some happy event by dancing to age-old tunes performed by their local, likely untutored, but effective, musicians. The sense of freedom of spirit within the music is captivating. *Slavonic Dances* was sculpted according to well-established musical-dance forms, including the furiant, polka, and dumka, but the Bohemian-sounding folk tunes were, for the most part, original to Dvořák rather than based on actual folk songs.

Dvořák was not alone in using folk songs for inspiration. Many composers have used them, completely or suggestively. Use of folk tunes helps identify composers' particular cultural or ethnic origins. For example, Ralph Vaughan Williams traveled around England, copying folk songs from many communities, for he recognized that the songs were in danger of being lost. Not content merely to record them for posterity, Vaughan Williams incorporated many of the tunes into his compositions, not the least of which was his *English Folk Song Suite*. Songs from the Cotswolds led to *Shepherds of the Delectable Mountains*. This work is mentioned because Robert Stradling (1998) has nicely established that Severnside (in the west of England, the region focused on the city of Gloucester with the Cotswolds as a southeastern border) was important for the development of British music, given that a fair number of composers either came from that region or visited it for thematic material.

Zoltán Kodály and Béla Bartok (1881–1945) also saved numerous folk songs, and they too used some of them—rewritten in their own inventive, respective, personal styles—to create an appreciation for their home landscape, in their case a lively Hungarian one, filled with energy and a sense of soul. These men worked for months gathering folk tunes in villages all over Hungary. Bartok accomplished a similar task in Romania, and Kodály did likewise in Transylvania. Bartok thereafter wrote *Hungarian Sketches* (1931), *Hungarian Peasant Songs* (1933), *Transylvanian Dances* (1931), and *Romanian Dances* (1911), while Kodály wrote *Galánta Dances* (1933) and *Peacock Variations* (1939) based on Hungarian sources and *Dances of Marosszék* (1929) using Transylvanian material. Johannes Brahms based some of his work on folk tunes, including especially his *Hungarian Dances* (1873). It is sobering to think that sounds from Hungarian, Romanian, English, and other cultural landscapes might have been lost forever due to rapidly changing societal circumstances if it had not been for certain composers and music educators who recognized the urgency to record the music before it was lost.

Dances from folk tunes abound in orchestral music. We can listen to orchestral dances (as musical geographies) that bring forth vivid images of people dancing in countries other than just England and Hungary. These include, for example, Poland (Henri Wieniawski's *Mazurka-Kujawiak* and Alexandre Tansman's *Kujawiak Mazurka*), France (Georges Bizet's *Danse provençal*, a joyous country dance as peasants in festive regalia make merry on the village green), Switzerland (Ernest Bloch's *Rustic Dances* from his *Concerto Grosso for String Orchestra and Piano*), Russia (Anatol Liadoff's *Russian Folk Songs and Dances*), Norway (Edvard Grieg's *Symphonic Dances on Norwegian Themes*), Spain (Enrique Granados's *Spanish Dances* and Isaac Albéniz's *Tango*), and China (Guo Wenjing's *She Huo* [1991] and *Melodies of West Yunnan* [1993]). Many other works could be cited. Worth mentioning also are Strauss's popular dance tunes for Vienna's high society, including *Eljen, A Magyar Hungarian Polka*, the descriptive *Thunder and Lightening Polka* (timpani as the thunder, cymbals and orchestral chords as the lightning), and *Pleasure Train*, a happy work with delightful percussive rhythms (with a piccolo for the station master's whistle and irregular beats for the train's going over joints between rails). The image of lively, whirling dance scenes in the various rural and, at least with Strauss, urban cultural landscapes is vivid. More will be noted on the use of folk tunes in chapter 5.

WOODLANDS AND SOME CREATURES THEREIN

References to folk songs are not the only way to refer to specific landscapes. For example, the New Zealander Gareth Farr (1968–) wrote *Waipoua*, in which he represents a New Zealand forest landscape with music that has been described as with the spirit of touching restraint. A work with a quite different character, Tōru Takemitsu's *Tree Line* (1988) for chamber orchestra, refers to someone walking beside a row of luxurious acacia trees located near the composer's villa in the mountains of Japan. *Elfinspiel* (1841) by Franz Berwald is a remarkable evocation of a Swedish landscape. We hear his elves dancing first on the violins, then jumping over to the flutes, before taking a larger jump to the basses, only to chase one another to the clarinets and oboes and, finally, to the bassoon. The composer's sister claimed his work contained "the ethereal essence of Nature." And a Jean Sibelius tone poem, *Skogsrået* (*The Wood Nymph*), is a delicious delight to the ear. As a boy, Sibelius enjoyed walking in the woods in search of fairies and goblins (Hill 1947, p. 10). Lest you dismiss my inclusion of these topics under the title Specific and Generalized, I should admit to having introduced my children, when they were young, to small patches of moss in one of Vermont's woodland clearances where, unfortunately, we always arrived too late to see the fairies and elves. But my children firmly believed they lived there. Charming representations of woodland creatures are also to be found in "Invocation and Dance of the

Will-o'-the Wisps" and "Dance of the Sylphs," selections from *The Damnation of Faust* by Berlioz. Boris Asaf'yev's *The Enchanted Lake* (1909) offers an enchanting fairy scene for orchestra that was inspired by *Kalevala* and the composer's visits to a lake at his wife's estate.

For sounds of humans altering the landscape, we can turn to *The Song of the Forest* by Dmitry Shostakovich (1906–1975). This work celebrates Stalin as "the great gardener" of the vast Russian steppes: Stalin's regime tried to remake vast blocks of the steppes into forest lands as part of the 1948 Great Plan for the Transformation of Nature (see A. Burke 1956). Responding to the dictates of the state, in 1949 Shostakovich wrote this huge work for choir and orchestra, which offers vivid encouragement for the planting of trees. The sections are titled "When the War Was Over," "We'll Dress Our Motherland in Trees," "Remembrance of the Past," "Pioneers Are Planting the Forests," "O Rise to Great Deeds All You People," "A Walk in the Future," and, derived from a nationalistic Russian folk song, "Glory." Of the numerous other composers to write about forest scenes on a much more intimate scale, we can note, as examples, Russian Alexander Glazunov (*The Forest*) and Frenchman Vincent Indy (*The Enchanted Forest*), each of whom wrote in a voice suggestive of the composers' respective places of birth and education. For sounds totally different from those created by the composers just identified, we can note Chinese composer Ma Xiang-hua's music for its use of the Erhu and a traditional Chinese orchestra to produce delightful—and distinctly Chinese—melodic tunes, including *The Grapes Are Ripe* and *Red Plum Capriccio*. Ma has also composed works for Western orchestras, including his Erhu Rhapsody no. 1. Mention of some of Ma's compositions leads nicely to the next section.

SOME CHINESE LANDSCAPES

Using explorations in timbre based on traditional Chinese melodic patterns, but for a Western orchestra, Chinese composer Chou Wen-chung (1923–) creates a lovely impression of Chinese landscapes in *Landscapes* (1949), from which one gets the sense of looking at a miniature Chinese landscape painting, so fine is the detail (see Chou 1998). And Guo Wenjing (1956–) expresses in music the beautiful and lively humanized landscape of Sichuan by skillfully drawing on traditional Chinese percussion, folk music from Sichuan, the style of singing found in Sichuan opera, and elements and instruments from Western music. His musical landscape is truly alive, even down to the representation of the cries of Yangtze river boatmen. Guo drew on folk tunes when he composed *Suspended Ancient Coffins on the Cliffs in Sichuan* (1983) about a particular human adaptation to cliffs that are massive in height and sheerness. Guo expresses an indebtedness to Bartok (from Hungary and Romania) and Krzysztof Penderecki (from Poland), though his own voice, his personal autograph, is clear. Guo's distinctive mixing of Chinese and Western influences (which he learned as a student in Moscow) follows the example of such composers as Alexander Borodin and Mussorgsky, who drew from East and West while, in their cases, creating music that was distinctly Russian.

British/Chinese Vanessa-Mae's *Happy Valley: The 1997 Re-Unification Overture for Violin, Orchestra and Chorus* is another example of mixed styles. ("Happy Valley" refers to the famous horse racecourse in Hong Kong, where a celebration occurred when Hong Kong reverted to Chinese control.) East and West meet well in her composition, a work that builds from the late Chinese leader Deng Hsiao Ping's concept of postreunification Hong Kong within China (i.e., one country, two systems), which, he was sure, meant, "The horses will keep on running, and the dancers—they will keep on dancing." The work is fascinating for its amalgam of styles. Lower strings and rumbling timpani suggest a threat (huge China), but then a single violin protests and grows in intensity, as if to say Hong Kong will not give up its sense of self. A racelike rhythm takes over, perhaps representing progress and the onset of a new future. Over this, the violin begins to sing a sad, yet sweet, theme, as if to say that the people of Hong Kong are ready to be conciliatory as they work together with the Chinese authorities to bring about a fresh, new approach to meeting the future. The work is filled with references to "old" China, as when a beautiful folk song is "sung" against a tattoo beaten on traditional drums. Military references abound, yet the rhythm of galloping horses and the singing violin solos suggest that the fiery spirit of the people of Hong Kong will survive in the new politico-administrative arrangement and that, indeed, "The horses will keep on running, and the dancers—they will keep on dancing." Some will like Vanessa-Mae's score; others will dislike it. That bias can influence how we react to music should not be surprising. The divide between likes and dislikes seem to be more clearly evident when consideration is given to music concerned with city landscapes.

CITY LANDSCAPES

The hustle and bustle of cities can be infectious, for those who like them. For example, Scottish composer Cedric Thorpe Davie (1913–1983) creates a lovely musical image of Edinburgh's Royal Mile as we join him, walking jauntily along the cobbled, historic mile-long route from Holyrood Castle (the royal residence), past the house of the famous Scottish reformer John Knox and St. Giles Cathedral, to Edinburgh Castle. *Royal*

Mile was composed in 1952 as a gift to Queen Elizabeth II on her coronation day.

Paris is one of the world's great cities. A well-known representation of that city is George Gershwin's *An American in Paris* (1928). A chirpy opening scene magnificently captures and conveys the image of pedestrians walking briskly and cars rushing hither and yon, with horns honking. For those who have been in Paris, the image is perfect. Gershwin's purpose was to portray the impression of an American visitor in Paris, strolling about the city while listening to street noises and absorbing the French atmosphere. The joyous opening flows into a rich blues section with a strong rhythmic background, as we imagine the visitor having a few drinks in a café and succumbing to, in Gershwin's words, "a spasm of homesickness." The bubbling exuberance of the opening returns, suggesting that the person is over his homesickness and, thus, is now alert to Parisian life, with its distinctive atmosphere and street noises.

The Paris of the English composer Frederick Delius has a different feel to it, but, of course, whereas Gershwin had a jazz and stage-show interest, Delius had a concern for delicate tone coloring. Delius's music also reflects something of his rather spirited bohemian lifestyle while in Paris. His *Paris, a Night Piece* (subtitled *The Song of a Great City*), composed in 1899–1900, describes, Delius wrote, his "personal record of feelings engendered by the city." Sometimes referred to as a miniature tone poem, the work incorporates melodic material that Delius heard in the streets as he walked about. He also wrote some lovely landscape music referring to other places: England—*Brigg Fair* (1907), *In a Summer Garden* (1908), *Summer Night on the River* (1911), and *On Hearing the First Cuckoo in the Spring* (1912); the United States—*Florida Suite* (1886–1887), *Hiawatha* (1888), and *Appalachia* (1929); and Monaco, specifically Monte Carlo—*Sur les cimes* (1893). Also of note, Scandinavian composer Johann Severin Svendsen (1840–1911) wrote a cheery *Carnival in Paris* (1872).

If an American, an Englishman, and a Norwegian can write about Paris, then a Frenchman can write about London, as Claude Debussy did in "Nuages," one of his *Nocturnes* (1897–1899). His London is considerably calmer than Gershwin's Paris. Slow in movement is perhaps the best way to describe Debussy's musical recollection of London and its cloud-filled sky. He was inspired by both his visit to the city and by James Abbott McNeill Whistler's paintings of London. A slight tension is in the air, as we imagine Debussy standing on a bridge over the Thames. A lugubrious theme suggests that a boat is slowly wending its way upstream while ever-changing clouds move overhead. Deep basses and a quiet timpani roll subtly tell of a storm, but since it is in the distance, the use of woodwinds reminds us that life is ongoing nearer to hand. In the second part, "Fêtes," in contrast, there is a burst of energy as we witness a lively, triumphant parade in Paris. Debussy either missed seeing the Royal Horse Guards in London or else he was not impressed by British imperialistic celebrations, favoring those in Paris instead! In the third section of *Nocturnes*, Debussy returns to the sea in "Sirènes" (see p. 58). Other compositions about London include Edward Elgar's *Cockaigne Overture, In London Town* (1901); Eric Coates's several works, including *London Calling, London Bridge*, and *Knightsbridge March*; and Percy Grainger's *Handel in the Strand*, a lighthearted representation of the hustle and bustle of people during a day along one of London's busiest streets.

Symphony no. 2, *London* (1913), is the major musical study of London. Vaughan Williams lived in a flat—an apartment in North American lingo—and from it he could see across the rooftops and their chimneys to St. Paul's Cathedral. Hugh Ottaway, (1972, p. 20) calls the work "large hearted music," while Vaughan Williams suggested the work should have been called "A Symphony by a Londoner." The work is a good example of musical geography for there are many small representations of the city, as in the first movement's vigorous *allegro*, which suggests the noise and hurry of the crowds. The movement also refers to the bells of Westminster, skillfully provided by the harpist. Vaughan Williams identified the second (slow) movement as Bloomsbury Square on a November afternoon, with a reference to the street merchants' cry of "sweet lavender." The scherzo has the listener at night standing on the Embankment, near Parliament, near enough to the Strand that distant sounds can be heard around the hotels, while on the opposite shore, people happy in their ventures fill the brightly lit, crowded streets. The final movement has a march rhythm, which, at first, is solemn, but then the main theme returns in an energetic manner. Upon hearing this music, one might think of soldiers marching, possibly with royalty coming along behind, but Vaughan Williams had something quite different in mind, namely, a hunger march. Near the end of the work, the noise and hurry of the first movement returns, though this time it is quieter, and Westminster's chimes can be hard once more.

Many other British cities have "been composed," but let me identify but one more, actually ten more, for their importance is significant. Rather than being descriptive of the city of Glasgow, Peter Maxwell Davies recently wrote the ten *Strathclyde* concertos, named for and financially supported by the Strathclyde Regional Council (with Glasgow at its core), with the intended soloists being the leading members of the Scottish Chamber Orchestra in Glasgow. Concerto no. 10 is a concerto for orchestra, rather than for one soloist. No other city seems to have been so favored. Each of the concertos has a different character, thus mirroring what is found in environmental perceptual research, namely, that people will see and believe different things in exactly the same landscape, even to the point of attributing values to landscape that are contrasting and even contrary in nature.

THE ETERNAL CITY

Historical and (then) contemporary aspects of Rome are portrayed in three grandiose "Roman" symphonic poems by Ottorino Respighi (1879–1936) in which he skillfully uses popular songs, brief selections of some old music, and his own vivid representations of the landscape. Each of the three symphonic poems has four movements. Respighi was explicit in his representations of Rome: he gave titles to each of the works, and he wrote detailed notes to describe the twelve movements, paraphrased here with additional comments.

Fontane di Roma (*The Fountains of Rome*) (1916) was inspired by four of Rome's many fountains. People are attracted to the fountains as much for the age-old practice of obtaining water as for a desire to look at the fountains' sculptured embellishments. Locals may take the architecture for granted; their focus is on getting water. Tourists, in contrast, look closely at them and often leave coins in the water in the hope that something good will happen to them, hence Sammy Cohen's Oscar-winning song "Three Coins in the Fountain" (from the 1954 movie of the same name, in which three women make wishes for romance at the Fountain of Trevi). The first movement, "The Fountain of Valle Giulia at Dawn," is quietly pastoral in character. (Not all European cities in the past were like cities of today with strict rules against urban sprawl; hence, until the early 1900s, pastoral activities could be carried out within the bounds of the city of Rome.) Sheep pass by before they and their shepherd disappear into the mist of the Roman dawn. This peaceful scene is rudely replaced by the joyful sound of birds and trumpets during the second movement, "The Triton Fountain at Dawn," in which tritons and nymphs seem to play together, run around, and dance in the spouting jets of water. The third movement, "The Fountain of Trevi at Midday," is represented by a triumphant tune on the cor anglais, echoed on trumpets, which introduces King Neptune on his seahorse-drawn chariot. Tritons and sirens follow, but all too soon, the procession fades and disappears into the distance. The final movement is "The Villa Medici Fountain at Sunset." A sad theme conveys a sense of nostalgia as the rippling water moves slowly across the pool that surrounds the fountain. From Respighi's music, one senses that the fountains are lovely to behold.

The second symphonic poem, *I Pini di Roma* (*The Pines of Rome*) (1924), presents four other visions of Rome. "I Pini di Villa Borghese" (Pines of the Villa Borghese) comes first; a delightful vitality leads us to imagine merry children's games. They play, laugh, dance, and mimic soldiers by both marching and fighting. The scene is dominated by huge trees that breathe with memories of the past. In being aware of the children, it is as if Respighi is saying that these very children will one day make their mark on history. Next, in "I Pini presso una catacomba" (The Pines near a Catacomb), a feeling of somber heaviness prevails, not simply because of the locale but also because the scene can barely be seen in the darkness of the shade provided by the pines. In that shade is the entrance to a catacomb. Muted strings and horns suggest mystery and gloominess, followed by an orchestral echo from the past. From the depths of the ground arises a sorrowful chant, like a solemn hymn briefly stated before mysteriously fading away. Chanting worshipers quietly and eerily sing—a wordless utterance in the lower strings—before a distant trumpet sounds. Next, in "I Pini di Gianicolo" (The Pines of the Janiculum), Respighi takes us to the famous Janiculum, a hill in Rome thought to lift up to the heavens. Under a full moon, the pine trees stand in bold profile as a nightingale sings against shimmering strings and a harp representing the night mist rising over Rome and lying languorously over the hillsides. Finally, there is "I Pini della Via Appia" (The Pines of the Appian Way). The scene shifts to a misty dawn on the Appian Way, where Respighi's imagination conjures up a vision of ancient glories. The scene opens quietly, yet undergirded by the threatening, steady beat of the timpani. Roman trumpets blare (often played from a high or rear balcony in the performance space), and into the brilliant light of the newly risen sun, with the timpani and associated heavy brass getting louder and louder, the army of the consul marches proudly toward the Via Sacre. Then, amid a rapturous crescendo, the soldiers triumphantly mount the Capitoline Hill. From playful children to a doleful catacomb, the silhouetted pines at night, and the proud army marching, the grandeur of Rome is pronounced.

Feste romane (Roman Festivals) (1928) is Respighi's third huge descriptive work. Filled with energy and verve, it depicts four of the city's festivals. The first movement, "Circenses" (Circuses), opens to a menacing sky centered over the crowded Circus Maximus (the ancient Roman circus that still stands today as a magnificent ruin). Despite the weather, everyone is there to celebrate the holiday. What better to do than go see Christians and others pitted against hungry lions! "Ave!" people call out when Caesar appears. Their cries turn to jeers when the huge iron doors in the circus are unlocked and swung open. A religious chant is heard as the Christian martyrs emerge from the holding cells, only to be interrupted periodically by howling of the animals. The anticipatory crowd noises mix with the animals' roars, and, together, they replace the martyr's chant, reaching a high pitch when there occurs the frenzied killing of the martyrs. The screams suddenly fold into a series of full-orchestral chords that seem to signal the final, brutal deaths. Immediately, there follows a sudden diminuendo and a grave, short, final statement that seems to signal the crowds' realization that there is no triumph. The second festival, "Il Giubileo" (The Jubilee), celebrating Easter in the Holy Year, opens with medieval pilgrims plodding along a country road, singing their prayers. Perhaps they are hauling large crosses, hence, their fatigue, or else they have come from afar. But Rome then appears in the distance, so they quicken their pace, and as soon as they get close to the city, they joyously sing a hymn of jubilation, joined by the city's church bells. "L'Ottobrata" (an October festival in the Castelli region of the Roman vineyards) is quieter and more reflective than the other movements. A romantic serenade is heard in the mild evening after a day in the vineyard. The final festival, "La Befana" (Epiphany), takes place on the eve of Epiphany in the Piazza Navona, where a loud clamor holds one's attention. We hear Respighi's characteristic use of the trumpet playing an old peasant tune with a rhythm that is quite catchy, plus bits of saltarello, a popular song of the inebriated, which, it would seem, includes the refrain "Let us pass, we are Romans." Respighi's three huge, multisectioned compositions serve to give a life to the historical city of Rome that is at once beguiling and, yet, also stunning.

NORTH AMERICAN CITIES

If we travel west by plane out over the Atlantic, we will reach a city in music that could only be New York. The New York of Leonard Bernstein's (1918–1990) *West Side Story*'s finger-snapping pulse is unlike any other city. Set in the west section of Manhattan Island, home of competing Puerto Rican gangs, the music gives one a mental image of jagged tension in anticipation of an inner-city drama. (Interestingly and strangely, the community used by Bernstein as his inspirational setting has since been removed, and the site now houses the Lincoln Center for the Performing Arts, home of the New York Philharmonic—Bernstein's orchestra from 1958 to 1969.) Based on *Romeo and Juliet* and coupled with Bernstein's intimate knowledge of the city, the resulting score is filled with fascinating sounds that create marvelous images, right up to the scored police whistle and siren. Originally written as a musical (1957), it was remade as a movie (1961), and it was scored as an orchestral suite, *Symphonic Dances from West Side Story* (1960). The mix of styles and rhythms could only be New York: blues and jazz are fused with "American modern," as it has been called, as the music relates the story of two lovers from different teenage street gangs. They have to contend with the continuing reality of tension and hatred between the gangs until, ultimately, the young man is killed as a result of the enmity.

In stark contrast to Bernstein's brilliantly alive city of the 1950s, Aaron Copland's (1900–1990) *The Quiet City* (1924) is a rather mournful work for chamber orchestra comprising strings, cor anglais, and trumpet. The high strings open the work, with plucked lower strings suggesting an underlying tension. The trumpet takes over, underlain by strings, and, with a slightly blues feel, expresses a mixture of pain and sorrow. The interplay continues, builds to a semiclimax by the strings, and then the cor anglais assumes the trumpet's role. The trumpet is not to be pushed aside, however, and it reasserts itself. Rather than compete, the two instruments play together in a fascinating, interwoven manner, accompanied by the strings. Though the work is called *The Quiet City*, it is at times loud, though never raucous. The sounds convey the sense that it is late at night: the strings represent the generally hushed, ongoing movement of people down on the street and in their apartments, while the trumpet and sometimes the cor anglais represent a person who, perhaps standing on a high roof top or at the window of an upper-floor apartment, looks out over the wonderful city he loves living in. He is alone, yet is not afraid to share his feelings with the world, though, perhaps because most people are asleep, few hear him. As a result, the work ends with a feeling of melancholy as the cor anglais expresses a final sigh.

Charles Ives (1874–1954) was a great descriptor in music of landscapes he knew well. One such is located in New York. While living in that city, he lived opposite Central Park. The park at night was a quiet place, but the streets beyond the park teemed with activity. This contrast he put in *Central Park in the Dark* (1906), written before the sounds of the automobile, radio, and TV, all of which, of course, were later to have a marked impact on the city's soundscape. He prefaced his score, paraphrased here. The work opens with a ten-bar repeated phrase in the strings that suggests, in Ives's words, "silent darkness." There is an intensity to the strings' mysteriously beautiful music (joined by clarinet and timpani, periodically interrupted by a trombone) that gradually increases in both volume and implied tension, until a jaggedness emerges to reveal, in the distance, in the city beyond the park, a place brimming with rambunctious life: the casino over the pond, street singers coming up from the subway, a street parade, newsboys calling out, pianolos playing in apartment houses, a street car, a street band, a fire engine, a cab horse running away, and more. Amid loud, jagged dissonances, we hear "Hello My Baby," a popular ragtime song of the day, which splinters the night, but after a huge burst of sound (marked *fff*) the music fades to a sense of quiet once more, as the view of the city is again focused on the park and its "silent darkness" in the near foreground.

A lesser known American composer and conductor, Werner Janssen (1899–1990), provided a quite different view of New York in his long neglected *New Year's Eve in New York, Symphonic Poem for Symphony Orchestra and Jazz Band*. The work's restless rhythms "suggest the aimless wanderings of the crowds along Broadway" as everyone awaits midnight. Various scenes are evoked, some with fire sirens and car horns, others with paper horns and rattles, until the excitement mounts as midnight approaches. In Charles O'Connell's (1941, p. 294) words, "the clock strikes twelve and pandemonium reigns. Here the jazz band is introduced, and the symphonic poem temporarily becomes a modern *concerto grosso*." O'Connell did not like the work and, indeed, was dismissive of Janssen's attempt to bring jazz into "serious music," though he did acknowledge that references to jazz might introduce some "novelties."

Skyscrapers (1923–1924) by John Alden Carpenter (1876–1951) was written first as a ballet and then rewritten as an orchestral suite. "Inescapable rhythms" of industry, construction, and, perhaps, office work, as well as the general haste and confusion of city life, are all present. The massiveness and utility of the skyscraper are implied by the "tremendous chords in full orchestra" (O'Connell 1941, pp. 179–81). New York is also a city of play, as evidenced in *Symphony for the Brooklyn Dodgers* by Robert Russell Bennett (1894–1981), written when the team was on its way to the 1941 baseball championship.

Some New Yorkers go for vacations to Cape Cod, Massachusetts; hence, the landscape and activities in Franz Bornshein's *Cape Cod Impressions* will be appreciated. Many other cities in the United States have been the focus of compositions, not least Bennett's *Charleston Rhapsody* and *Hollywood*, Ferde Grofé's *Hollywood Suite* and *San Francisco Suite*, and Meredith Wilson's Symphony no. 1, *San Francisco*. Grofé also composed a work for a once important Western town, *Virginia City: Requiem for a Ghost Town*.

SPECIFIC LANDSCAPE REFERENCES: A NEW ENGLAND INTERLUDE

Mention of Cape Cod just above, located as it is in New England, brings me again to Charles Ives. *Three Places in New England* (completed in 1914) is tied to specific urban locations. The first section of the work describes a scene engraved on the Boston Common's Augustus Saint-Gaudens monument. The work is entitled "The 'St Gaudens' in Boston Common (of Col. Shaw and His Colored Regiment)." (The story of Col. Robert Gould Shaw and his shared fate with the first unit of African American volunteer soldiers, many of them former slaves, during the American Civil War is told in *Glory*, a 1989 movie.) The work opens in dancelike fashion, heard quietly, as if from far away—perhaps in time as in space—and then moves to a series of almost marchlike rhythms interspersed with more reflective passages, representing the soldiers resolutely marching and singing their songs. Ives prefaced his score with a poem, which reads, in part,

> Moving,—Marching—Faces of Souls!
> Marked by generations of pain,
> Part-freers of a Destiny,
> Slowly, restlessly—swaying us on with you
> Towards other Freedoms!

As the drum picks up the beat, Ives directed, "From here on, though with animation, still slowly and rather evenly. . . . The drum seems to follow the feet, rather than the feet the drum" (Ives, quoted in Feder 1992, p. 235).

The second part of the work, "Putnam's Camp, Redding, Connecticut," tells a story from the Revolutionary period. Near Redding Center is a small park preserved as a memorial to Gen. Israel Putnam's soldiers, who wintered there in 1778–1779. As Ives wrote, "Long rows of stone camp fireplaces still remain to stir a child's imagination. The hardship which the soldiers endured and the agitation of a few hotheads to break camp and march to the Hartford Assembly [the seat of government], is a part of Redding history." The scene in the music is not of the soldiers but of children in the park, for a July Fourth picnic. As a child wanders from the main gathering in hopes of catching a glimpse of the old soldiers, the tunes of the band and songs of the children in the park grow fainter. A tall woman is seen, seeming to stand over the trees on the crest of the hill. She reminds the child of a picture he has of the Goddess of Liberty. Her face is sorrowful. She pleads with the soldiers to not forget their cause and the great sacrifices they have to make for it. But they march out of camp with fife and drum to a popular tune. Suddenly, a new national note is heard: the general is coming. The soldiers turn back and cheer. The little boy awakes. He once again hears the children's songs and runs down past the monument to listen to the band and join in the games and dances. As we listen to the music, Ives's band has a steady beat, until it reaches a quiet section that swells into a section punctuated by brass and offbeat percussive sounds that jar. Eventually, a quieter marching tune returns, but it too is replaced by a full-orchestra discordant march. Interestingly, one of the tunes we hear is "The British Grenadiers," for it was popular among the American soldiers too, but with changed words. (In another work, *"Country Band" March,* Ives presents a parody of a New England town band.)

These first two parts of *Three Places in New England* represent specific landscape features and the stories associated with them. Ives was noticeably tied to landscape references in his music (Cooney 1995). In the third part of the suite, he turned more generally to a landscape he knew well.

"The Housatonic" is reflective, representing a walk Ives and his wife took one Sunday summer morning in the meadows near Stockbridge, Connecticut, with running water, church bell, and a short hymn that grows quickly and is then released. He wrote, "The mist had not entirely left the river bed, and the colors, the running water, the banks and the elm trees were something that one would always remember" (cited in Ives 1972, p. 87). The Stockbridge landscape must have been beautiful indeed.

MOVING ON

Moving north we arrive in Montreal, as described by George Fiala (1922–). His symphonic suite *Montreal* is somewhat suggestive of Gershwin's *An American in Paris*. One section of the suite, "Metro," is innovatively descriptive in its suggestion of the rhythm and tension of a subway station. West to Toronto, we come to John Beckwith's *Place of Meeting* (1967) for tenor, choir, and orchestra, a jagged work that deals with the problems of human fulfillment in the modern world. Beckwith (1927–) saw Toronto as a city of branch plants for multinational corporations, leaving Torontonians as victims struggling for survival (David Parsons, personal communication, 1986). Music from Toronto likely will change markedly since UNESCO now holds that city to be the most multinational and multicultural city in the world, with more than 51 percent of the population having been born outside the country. Another city undergoing rapid change, Johannesburg, South Africa, is celebrated in music by Sir William Walton. His chirpy *Johannesburg Festival Overture* (1956) was written in honor of the seventieth anniversary of the founding of the city. Walton had some African music sent to him in England, and he incorporated some of its themes into the work; however, Walton's "imperial" style is never lost.

Heading to New Zealand, we find that many composers have been inspired by and written about the landscape, including urban scenes. For example, Larry Pruden's (1925–1982) *Harbour Nocturne* (1956) is a delicate portrayal of the impressive harbor at Wellington, the capital city. Pruden also wrote short works entitled *Westland*, *Taranaki*, and *Akaroa*, named for regions of the country. Auckland, the country's largest metropolitan center, "The City of Sails," has been captured in *Vistas* (1966), five musical images by John Wells (1948–), and in *Auckland!* a fanfare by Christopher Blake (1949–). John Ritchie (1921–) composed a lively concert overture for full symphony orchestra entitled *Papanui Road* (1987), which explores "the bustle, the vitality and the peace" of one of Christchurch's busiest arteries. Christchurch is on the Canterbury Plain, with the ocean to the east and the striking chain of mountains, the Southern Alps, in the distance to the west. Christopher Blake's *Leaving the Plains of Canterbury* (1979) for chamber orchestra creates a soundscape of the setting. Blake's *Night Walking with the Great Salter* (1982) explores the world of small-town New Zealand, with its "black and dangerous undercurrents depicted in atmospheric and unpredictable music" (Christopher Blake, Centre for New Zealand Music).

Moving to Southeast Asia, we can enjoy *Rush Hour in Hong Kong*, by Abram Chasins, but since he wrote the work in 1935, I daresay rush hour in that populous city has far more "rush"—or perhaps "slow"—to it now than Chasins conveyed. Tan Dun composed *Symphony 1997*, a large orchestral work for the day Hong Kong was returned by the British to China. It is hard to place the work centrally in Hong Kong, other than, perhaps, by accepting that Hong Kong is now a cosmopolitan society, despite being Chinese, that perhaps no one style of music clearly fits.

Many other cities could be discussed in this manner. If we were to add brass band music, the number of items would go on and on, for hundreds of works are named for such cities as Dunedin, Invercargill, and Wellington. Adding U.S. concert bands, *Lincoln Square March*, *Manhattan Beach*, and *Washington Post* come to mind, though the last march mentioned is named for the newspaper, not the city. And still more works could be cited if pipe band scores were added, including *Bonnie Dundee* and *Inverness Gathering*. In contrast to music dedicated to cities, there are also compositions on broader regions and physical features, such as *Loch Lomond*, *Bonnie Galloway*, and *Mist Covered Mountains* (pipes), on the U.S. South in *King Cotton*, and on states such as *Michigan* and *Wisconsin* (concert band), but these works are not descriptive. Let us now return to orchestral music.

INDUSTRIAL LANDSCAPES

The list of orchestral music celebrating industry is surprisingly short, apart from many works written by composers in the Soviet Union in response to the Stalinist demand that music should conform to a "constructivist" or "formalist" style (to use the terms Soviet authorities used). Soviet music had to glorify the workers and the state. Aleksandr Mosolov, for example, wrote *Soviet Iron Foundry*, a noisy, but formless, attempt to create an image in music of a steel mill, with flaming forges and ceaseless activity. And Shostakovich's Symphony no. 2 (*October*), dedicated to the October Revolution of 1917, includes a factory whistle. Another Russian work, by Serge Prokofiev, a collection of six orchestral pieces extracted from a ballet entitled *The Age of Steel*, is now rarely, if ever, heard. The suite's sections are "Train of Men Carrying Provision Bags," "Sailor with Bracelet and Working Women," "Reconstruction of the Scenery," "The Factory," "The Hammers," and "Final Scene." A British reviewer commenting on the work's premier in the United Kingdom stated that Prokofiev has "a hard and steely style" and, interestingly, that "the percussion does not suffer from reticence."

"Industrial music" was not confined to the Soviet Union. For example, a once-heard work of similar ilk by American Harl McDonald (1899–1955), *Festival for the Workers* (1934) used tone pictures to convey a labor rally, held perhaps in Philadelphia where McDonald lived. The first movement, "Procession of the Workers," included the pulse of thousands of heavy-booted feet approaching from the distance under a melancholic bassoon line. "Dance of the Workers" followed; the rhythms of the dance seemingly reflected sadness as much as gaiety. The final movement, "Exaltation of the Workers," was destroyed by the composer! Another American composer, Ferde Grofé (1892–1972), wrote *Symphony of Steel* (1935), but it, too, has disappeared from the orchestral stage. Of a quite different ilk, English composer Eric Coates wrote *Calling All Workers* during World War II. It is still performed. The work was dedicated to "all workers" so that each one could "go to . . . work with a glad heart and . . . do that work with earnestness and good will." The constancy of the beat represents a factory filled with sewing machines.

Early in the last century, "Fordism" got underway with the mass production of cars on assembly lines, which in time led to "just on time" production. In the mid-1920s, the Ford Automobile Company used as a billboard advertisement, "The Ten-millionth Ford Is Now Serving Its Owner." Impressed by the advertisement and Arthur Honegger's *Pacific 231* (see next paragraph), New England composer Frederick Converse wrote *Flivver Ten Million: A Joyous Epic Inspired by the Familiar Legend "The Ten-millionth Ford Is Now Serving Its Owner,"* which received its premier performance in 1927. The work was decidedly programmatic, for the eight sections included "Dawn at Detroit" (sunrise, the city awakes), "The Call to Labor" (bells and factory whistles), "The Din of the Builders" (factory noises), "The Joy Riders—America's Frolic," and, pointedly, "The Collision—America's Tragedy." Another work inspired by cars was commissioned by the Toyota Corporation. Canadian composer David Warrack was asked to write an orchestral work for the opening of a new automotive assembly plant in Ontario. *Overture for a New Era* (2001) was later arranged for symphonic band. The work is rhythmic and pleasantly melodic. I had trouble imagining a car plant as I listened to the work, but the energy and drive does imply the celebration of "a new era," so that surely would have made the Japanese automobile company executives happy.

Without a doubt, the most industrial sound in music is that provided by French composer Arthur Honegger (1892–1955). What you see and experience on a regular basis may stimulate thought, if you are open to it. Such happened for Honegger, who, when a boy, always was excited when walking past the railway yards in Le Havre. In time, he became passionately in love with locomotives: "For me they are living beings whom I love as others love women or horses" (Honegger [1923] 1975, p. iii). When in his early thirties, he realistically expressed this love in music. *Pacific 231* (1923) premiered in Paris. He also wrote that he wanted to create "the visual expression and the physical sensation" of the change from the "quiet respiration of an engine in a state of immobility; [the initial] effort of moving," with whistle blowing and brakes screeching as they are let go, and the "progressive increase of speed, in order to pass from the 'lyric' to the pathetic state of an engine of 300 tons driven through the silence of the night at a mile a minute." The music expresses raw power as the machine starts to move, gains speed, and then races at full throttle. One critic of the premier performance called it a veritable hymn to the glory of speed, whereas another was both shocked by the brute force evoked in the music and depressed by its inhumanity. Yet another dismissed it as a "rather cheap imitation of a locomotive and its strident dissonances" (O'Connell 1935, p. 289). However, Olin Downes (1943, pp. 232–33) has the final word here: he feels the work conveys the "power and glory of motion" but also "laughter" from Honegger, who had "a deep-chested laugh." Further, Downes comments on the feat of slowing and stopping an engine after it has reached full throttle: it "is a crucial test for the engineer in the cab as it is for the composer at his score paper—the slowing up of the great mass that has gone hurtling through the darkness, and the stop, with the harsh, grinding harmonies of the last measures" (O. Downes 1943, p. 233). Honegger's powerful *Pacific 321* is, in effect, musical geography, without parallel.

CONCLUSION

This chapter has discussed a number of examples of music that relate to both specific places, or to features in them, and to generalized landscapes. That composers can give life to such places is remarkable in terms of the range and scope of their interpretations. One cannot think Honegger's locomotive was written by Beethoven, just as Beethoven's landscape could not have been written by Ives. As discussed earlier, each composer has his or her own voice, to be enjoyed for itself. Next we turn to imagined and mythic landscapes.

5

Imagined and Mythic Landscapes

Historical geographer Ralph Brown, in a richly developed examination of the creation of the North America settlement canvas, declared that people "are influenced quite as much by beliefs as by facts" (1948, p. 3). We have seen in the previous chapter that some composers responded to "facts" and wrote accordingly. In this chapter, I consider how some composers have responded to imagined landscapes, some in stories or inspired by paintings or, even, birds, while others are set within legendary and mythic formulations. Some of the latter find expression in nationalism. The resulting landscapes in music can have special meaning for the people with whom each composer shares an interwoven cultural and national identity. Still other landscapes in music refer to mythical places for the dead, for a discussion of which, see chapter 7.

THE FIRST PROGRAMMATIC SYMPHONY

Symphonie fantastique (1830) by Hector Berlioz (1803–1869) stands as one of the most remarkable bits of orchestral music ever written and also as the first completely programmatic symphony. Berlioz's hero was Ludwig van Beethoven, in part because in 1827 he heard a performance of Symphony no. 3, *Eroica*. He was well versed in the writings of Johann Wolfgang von Goethe, and he knew of William Shakespeare's writings, not least because of a touring company that performed *Hamlet*. The story goes that he was struck as much, if not more, by the leading actress as he was by the play. He supposedly wrote his symphony while he was under the delirious influence of imagined love for her. She knew nothing of him at the time. Further, the *Symphonie fantastique* premiered without her knowledge. They eventually got married but did not live happily. The following comments on the astonishingly inventive, five-part "tone poem" (a phrase not in use when Berlioz wrote his work) are based, in part, on Berlioz's own detailed description of his programmatic work.

A young musician of morbid sensibility and ardent imagination poisons himself with opium in a fit of amorous despair. He does not die but plunges into a deep sleep and has strange visions. His beloved becomes for him an endless melody that he finds and hears everywhere—an idée fixe (i.e., a recognizable statement that when repeated is instantly appreciated by the listener). There are five sections.

The idée fixe appears in the opening orchestral section, "Reveries," which includes an array of emotions, from melancholy and joy, to delirious anguish and jealous fury, but ends with loving tenderness. "A Ball" is set at a tumultuous party. Inspired by Beethoven's *Pastoral Symphony*, Berlioz next wrote a "Scene in the Country," which opens with an interchange between the cor anglais and oboe to suggest a delightful summer evening, with two shepherds playing a duet on their pipes amid lightly rustling trees. The idée fixe intrudes fleetingly, and the musician becomes agitated. One shepherd (cor anglais) resumes his melody, though the other shepherd does not reply. The sun sets, and then, in what was an innovation, four timpani convey the threatening sound of distant thunder (see p. 136). Quietness comes, and the musician is left by himself, alone and lonely. In the vivid imagery of "March to the Scaffold," the young musician dreams he has killed his beloved and is condemned to die. The procession to the scaffold is at times wild, even demonic (with massed brass and percussion), and at other times somber, with the steady timpani beat suggesting the tread of heavy feet. At the last moment, as the orchestral fatal stroke is about to happen, the idée fixe appears for an instant, played by the clarinet, like a last love-thought, and the axe falls. "Dream of a Witches' Sabbath" is set in the midst of a frightful group of witches, ghosts, and monsters of all kinds, who have come together for the musician's funeral. Strange noises, groans, ringing laughter, and shrieks fill the air. The idée fixe appears yet again, but it is now bereft of nobility and shyness, for it has become an ignoble, trivial, and grotesque dance tune. A diabolic orgy occurs, as does a burlesque parody of the plainsong "Dies irae" (which foretells the Day of Judgment). The latter melody and a frenzied witches' dance combine, and the work surges to a wild conclusion.

Berlioz's landscape imagery is startling. His use in each movement of the idée fixe was innovative, and his brilliant orchestration was decades ahead of its time; yet, interestingly, as noted above, he built on the recent past as expressed in Beethoven's work. Berlioz's color combinations, rhythms, and tones startled the audiences of his day, and as if that were not enough, he used an orchestra of far larger than hitherto standard size. The French critic François-Joseph Fétis felt that "Berlioz was trying the impossible by seeking to describe in music both physical objects and moral quality" (E. Downes 1976, p. 150) but that he did so successfully. Berlioz thus opened up a new way of approaching music, whereby imagination could fly free of convention and, through the use of both skillful composition and stunning effects, express more than "just" a musical idea. Programmatic music was thereby clearly given a stunning "launch," though it stimulated heated debate. The Viennese critic Eduard Hanslick, composer Johannes Brahms, and others felt that it was improper for music to tell a story and that "absolute music" had prime place. Berlioz, Franz Liszt, Richard Wagner, Felix Mendelssohn, and countless other composers, with their composing pens, established that music can sometimes be about something. Modest Mussorgsky (1839–1881) is another composer of this ilk, so let us now turn to one of his works.

WALKING AROUND AN ART GALLERY

Composers' imaginations have often been stirred by paintings. Modest Mussorgsky's *Pictures at an Exhibition* (1874) is loosely based on ten oil and watercolor paintings by Viktor Hartmann. Since Mussorgsky may never have seen the actual paintings, and since his musical creations are more likely make-believe scenes, what does that make the music? A double-double make-believe? In any case, Mussorgsky wrote a piano suite based on his understanding of the paintings. Maurice Ravel's ingenious orchestration saved the work from oblivion.

We have two landscapes to deal with, an outer space (the gallery) and an inner space (the scenes within each painting). The work begins with a bold and very Russian theme called "Promenade," played on solo trumpet. Subsequent variations of the promenade theme, as our visitor walks through the gallery, are played by bassoon and oboe or various other combinations of instruments, and sometimes by full orchestra. It is easy to imagine a gallery visitor, perhaps Mussorgsky himself, ambling in an uneven gait from one painting to another.

The first section, "Gnomus" (Gnome), represents a grotesque little person. It is portrayed by woodwinds, plucked strings, and muted brass playing descending scales. It is not known if the person is real or not, but one fancies not. Next, "Il vecchio castello" (The Old Castle) reveals a medieval castle set in an Italian landscape at night with a troubadour singing to his lady, which Mussorgsky portrays by using a bassoon to sing a wistful song, followed by an alto saxophone with a string accompaniment. "Tuleries" then portrays a famous Paris park where attending nurses go with children, who are chattering as they play: the sound is stereotypically musical geography. There follows "Bydlo" (the Polish word for a crude ox-drawn farm wagon), which represents cattle at Sandomierz, a village near Krakow, to which Mussorgsky, using a folk song, added the oxcart moving slowly and irregularly on its enormous wooden wheels. The next painting clearly tickled Mussorgsky's fancy, for his representation of it in "Ballet of the Unhatched Chickens" is a delight, with flute and oboe chirping, suggesting that the chicks are bouncing around and pecking as they dance inside their shells. Next comes "Samuel Goldenburg and Schmuyle," two Polish Jews. The rich man (a suave melody played by strings and woodwinds) converses with the poor man (a nervous thin-sounding trumpet). Their conversation gets entangled before concluding abruptly, the rich man dismissing the poor man. "Limoges—the Market Place" follows, in which a theme played in vivid color by strings and trumpets reveals women arguing in a small, French market town. Then, "Catacombs" (see also p. 150) draws upon the notion that Hartmann explored the catacombs of Paris and, using lantern light, painted what he saw. Mussorgsky marked this section *Cum mortuis in lingua mortua* (with the dead in a dead language). Solemn and churchly in character, the woodwinds "moan" above muted strings. Ascending scales reveal the movement upward from the underground to the next scene, "The Hut on Fowl's Legs" (sometimes referred to as "The Hut of Baba-Yaga"). Baba Yaga is a legendary Russian witch portrayed by a trumpet using a mute. She enjoys gathering human bones and pounding them to a convenient size with her pestle. To add some spice, Mussorgsky includes a lively witches' chase. Thereafter, the work concludes with the "Great Gate of Kiev," a gigantic musical image of this never-actualized, planned entrance to the city; so, the work clearly ends with an imagined landscape. Deep brass implies a major, high, turreted facade, but a mournful trumpet seems to suggest that the facade will never be. Even so, we are then led on as the music draws, in majestic fashion, upon the promenade theme and moves to a dynamic conclusion.

HOT DAYS IN ARID ENVIRONMENTS

Movement of animals across dry lands generally needs to be slow but purposeful, going from water source to water source. Such is the scene in a delightful work by Russian composer Alexander Borodin (1833–1887). The scene draws from his imagination, perhaps of caravans going from the Urals through the steppes to the distant parts of the Russian Empire. Distances were vast, populations encountered were not always friendly, and pressures on the travelers to trade their goods and get home as quickly as possible were considerable. Huge, featureless plains burned by the hot summer's sun were at times desperately hard to move through. The heat of the day was the worst, for tricks of the air caused mirages to be seen, which in turn raised false hopes among the travelers. Borodin captures the slow-moving monotony and hardship of such a trip magnificently in his *In the Steppes of Central Asia* (1880). The preface to this imagined "orchestral sketch" was provided by Borodin and is paraphrased here. Out of the silence, we hear the sounds of a peaceful Russian song, followed by the melancholy strains of Oriental melodies played as an almost monotonous wail on the oboe. The sound mounts as we hear the stamping of approaching horses and camels, and a large caravan comes into view. The hot sun beats down. With full trust in its protective escort of Russian soldiers, the caravan continues in a carefree mood. The long procession shambles past and off into the distance on its seemingly unending journey. The songs of the Russians and those of the Asiatic natives mingle in common harmony. The refrain curls over the steppe and then dies away in the distance; thence, nature is returned to its peaceful state, without human presence. We turn now to a nearby region.

A STORY OF MANY STORIES

Based on an old Arabic collection of tales called *The Thousand and One Nights*, Nikolay Andreyevich Rimsky-Korsakov's *Scheherazade* (1888) captures the mood of some southwest Asian cultural landscapes. It was not his intention to tell in detail the story that forms the basis for the work, but a quick note is appropriate. Sultan Shahryar is so angered that one of his numerous wives has cheated on him that he vows to achieve revenge by killing one wife each night. Princess Scheherazade knows of this, so when her turn comes, she starts to tell him a long tale, but as the sun rises she stops just short of the conclusion. The sultan, interested in the outcome, demands that the princess return the next night to continue the tale. Soon, however, she launches into another tale, ending it, too, just before the full story is revealed. She repeats this pattern until one thousand nights have passed. Each of her tales includes great wisdom and love and often is about the adventures of heroes. By the time all of the tales are told, the sultan has fallen in love with the princess, and so, he gives up his plan of revenge.

Scheherazade's thunderous opening is menacing, conveying the presence of the sultan. Soon thereafter, a solo violin, trembling and hesitant, yet also graceful, plays the Scheherazade theme against a harp background. Sinbad's story is told, at first calmly but then with the full fury of the seas. The second movement opens with the Scheherazade theme stronger than before. The bassoon solo gives us the jaunty Kalendar Prince theme, repeated on oboe and later on violins, woodwinds, and a solo horn. The pleasant sense of peace is rudely shaken as the scene becomes one of barbaric splendor, with brass fanfares and whirling colors brilliantly portrayed by the strings and woodwinds. The sense of splendor—and movement—ends with the mullah's call to the Muslim faithful. The third movement is lyrically romantic. The themes for the sultan and the princess are played tenderly, with an ethereal quality to the instruments' rich tone colors. A friend once told me that the beautiful interplay of colors reminded her of the sun catching the splendor of a peacock's tail. Movement four opens with the loud voice of the sultan, but Scheherazade diverts his anger with a description of a festival in Baghdad. The flute music greeting is soon joined by a variety of instruments that create a whirling joyous swell as Scheherazade tells of happy people milling around, sinuously weaving dancers, snake charmers, and the beauty of rare perfumes. The sultan's ominous theme returns, only to be quelled once again by Scheherazade's return, after so many months, to the Sinbad story, with his ship caught in a raging storm. Then, with a shattering smash, the ship crashes onto the rocks. The full tale now told, the sultan contemplates her achievement and speaks to her gently, and the work ends with Scheherazade's theme fading upward through a final, peaceful chord from the orchestra. If Rimsky-Korsakov in *Scheherazade* was a wonderful teller of tales, so too was Paul Dukas, to whom we now turn.

A SILLY APPRENTICE

The next work to be considered was based on a humorous poem by Goethe, which he, in turn, borrowed from an ancient Egyptian satiric tale. Paul Dukas (1865–1935), a native of Paris, is best remembered for *The Sorcerer's Apprentice* (1897). Dukas prefaced his score with Goethe's poem to be sure that no one missed the point of the music. The story is about a young apprentice to a philosopher, who learns all he can from his master. However, the latter will not tell the apprentice the spell needed to get a broom to do things, so the apprentice hides one day and overhears the master using the spell. When the master is next out, the apprentice gives the broom the three-syllable command and demands that it fetch water. The broom obeys. "Good," he says, and adds, "I want no more water, so be a broom again," but the broom ignores him and keeps on getting water. Soon, the room is overflowing. Frightened, he grabs an axe and chops the broom in two, hoping that the nightmarish situation will end, but it gets worse, for the two bits of broom rush faster and faster, bringing more and more water. The place is flooded when the philosopher returns. He quickly returns the broom to its original state, whereupon the apprentice sneaks out and never again sees the philosopher.

Dukas took this tale of sorcery and magic and retold it in music, using the orchestra's palate as a remarkable source for stunning humor. Muted strings quietly present a mysterious, hocus-pocus theme (that is evident throughout the work). A flurry of sound from the woodwinds represents the apprentice, while the master is at first represented by the bassoons. The master then leaves, and the woodwinds announce the magic spell (which quickly evolves into the major theme of the work). A sudden thwack on the timpani silences all, and we wonder what has happened. A low "grunt" from the orchestra reveals that the broom is moving. The grunts multiply and acquire an awkward rhythm—the broom (bassoons) is fetching water. The apprentice's theme returns, and a sharp climax suggests the chopping of the broom, followed by the even more rapid enunciation of the theme as the two brooms rush to get water. The helpless apprentice is amazed and increasingly alarmed. The main theme returns and mixes with the apprentice's theme in a hilarious manner, as if the instruments are laughing. The master reappears (loud, sustained blasts on trumpets and horns), and order is restored. The opening string theme returns but fades as the orchestra moves to the end in an explosion of comic relief. With the final four startling chords, surely Dukas himself is laughing.

These works by Berlioz, Mussorgsky, Borodin, Rimsky-Korsakov, and Dukas reveal emphatically that some composers are skilled storytellers. The images of landscapes (whether of the march to the gallows, the distant thunderstorm, the market in Limoges, the catacombs, a slow-moving caravan, the sultan's palace, the violent seascape in which Sinbad's ship helplessly flounders, the boisterous festival in Baghdad, or the scene in the philosopher's house) are remarkable, as are the ways in which the people in these landscapes are identified in music as varyingly fascinating, frightening, charming, loving, and mischievous. These several works are each so different in color and intent, yet all are superb examples of composers' skill at creating programmatic music, with distinct references to imagined people and their landscapes. Another composer, Richard Wagner, also created giant musical landscapes.

FOREST MURMURINGS AND TONE COLORS

Wagner's *Waldweben* (*Forest Murmurs*), from his opera *Siegfried* (1876) is profound in its representation of imagined nature. Low and almost inaudible murmuring by the strings becomes brighter in sound as muted violins join in to create a beautiful atmosphere that helps Siegfried recall times from his youth. The linking of elements in nature (sun, shifting light, wind in the trees) with Siegfried's recollections is marvelous, for the music reveals different levels of meaning. While Siegfried sings about a variety of issues, the orchestral music uses a theme that signals a subtle message about his origin. Birds sing, first the clarinet, then the flute. They tell Siegfried of a wonderful woman who is magically asleep beyond a ring of fire. The orchestra plays quiet slumber music, a suggestion of Loge's magic fire, and a prophecy using Siegfried's own call on the horn. Hearing the call, he asks for help, which comes in the form of a bird that leads him toward Brünnhilde's fire-bound rock. I assure you, the music sounds far better than this description reads, but operas do tend toward the bizarre in terms of their story lines!

Wagner was a genius at crafting wondrous tone colors to represent geographical phenomena, including a storm at sea, the trilling of many birds, and, of course, the clumping of the mythic giants as they approach. His huge dramatic cycle known as the *Ring of the Nibelung* contains several leitmotivs (a recognizable statement in music that, when repeated, is instantly appreciated by a listener for its representation of a person, an idea, a place, an emotion, or a situation in a musical drama) for such elements as Valhalla, the Rhine, Valkyries, water maidens, gold, and a sword. These leitmotivs are sometimes layered, one on top of another, though each is a character in its own right. Some of Wagner's "characters" are nonhuman. It is to stories that include other nonhuman characters that we now turn.

THE ANIMAL WORLD

Some works enjoyed by children refer to well-known stories or incorporate readily identifiable animals and birds. The most popular such work undoubtedly is Serge Prokofiev's *Peter and the Wolf* (1936). Prokofiev called this chamber orchestra work "a musical tale for children." The literal translation from Russian is "How little Peter fooled the wolf." At performances that use a narrator (reading Prokofiev's version of the famous tale), the various instruments are introduced before the work gets fully under way: Peter—strings; Bird—flute; Duck—oboe; Cat—clarinet; Grandfather—bassoon; Wolf—horns; and Hunters with guns—timpani and bass drum.

A much earlier work, Leopold Mozart's *Toy Symphony*, is a delight for its simplicity and humor. First and second strings, and basses, are joined by toy instruments: birdcalls (cuckoo, quail, and nightingale), a toy one-note trumpet, a toy drum, and a ratchet. There are three miniature movements, the last of which is repeated three times, each one faster than the last. When the superb Eszterházy musicians first performed this work, they chuckled so much they could not keep time. Another toy work is called *Machines and Dreams: A Toy Symphony* (1991). Composed by English composer Colin Matthews, it uses modern day "toys": toy pianos, bird whistles, a fishing rod/reel, computer sounds, a flyswatter (to silence the noisy bugs and birds), a car horn, bells, and much more. "It is surprisingly melodic and lush, and everyone has a great time—especially the orchestra" (Alan Best, personal communication, November 28, 2004).

Franz Joseph Haydn's Symphony no. 83 (1785), named *La poule* (*The Hen*), is another humorous work. It includes the representation of chickens cackling. Haydn drew upon the examples of others' when he composed, including Carlo Farina of the Court of Dresden, who, in the early 1600s, had imitative "cackling hens" and "barking dogs" in his *Capriccio stravagante* (1627). Antonio Vivaldi was another leader in this genre. As identified in chapter 2, his *The Four Seasons* (ca. 1730) includes bird songs, dogs, and insects. Some of these early scores had, at times, an almost carnival-like atmosphere. It is to a carnival of animals that we now turn.

Camille Saint-Saëns's (1835–1921) wonderful, tongue-in-cheek *Carnival of the Animals* (1886) is filled with humorous sounds and images. It calls for two pianists, a chamber orchestra, and, if the nutty text by Ogden Nash is used, a narrator who reads Nash's poems between each of the fourteen pieces of music. Saint-Saëns, as part of his joke, quoted other composers' material. To deflect any annoyance on their part, he also quoted one of his own works. The music makes best sense if the subject of each number is known. The work opens with the "Royal March of the Lion" (pianos and strings bow to the lion as he marches in and roars in menacing keyboard runs). Thereafter come, in order, hens and roosters (strings and pianos), a wild jackass (frantic racing up and down on the pianos), tortoises (two themes from Jacques Offenbach's *Orpheus in the Underworld* played by pianos and strings very slowly—at tortoise pace), elephants (a hilarious image created by the double basses in a quote from Berlioz's *The Damnation of Faust*, with the lugubrious elephants dancing the delicate *Dance of the Sylphs* and a tiny part of Mendelssohn's *A Midsummer Night's Dream*), kangaroos (bouncing around, on pianos), an aquarium (in which pianos, strings, and flute are the fish swimming), mules (a shrill hee-haw interlude on the strings), a cuckoo in the woods (clarinet with piano background), and another bird (a "flighty" flute fluttering here and there). Not shy at poking fun at pianists, perhaps because he was a superb pianist himself, Saint-Saëns moves into "The Pianists," in which he directs pianists to "imitate the gaucheries of a novice" as they rush up and down their scales. This short piece is a wonderful send-up by Saint-Saëns of serious pianists who, in turn, by performance tradition, alter tempi and make the work seem exceedingly difficult, even as they actually play it brilliantly. Thereafter come "Fossils" (xylophone, pianos, strings, flute, and clarinet in a delightful piece with references to five works, including "Twinkle, Twinkle, Little Star," Gioacchino Antonio Rossini's *The Barber of Seville* and his own *Danse macabre*) and "The Swan" (gracefully swimming around, represented by a lovely solo cello). In the grand finale, the various themes are brought together.

Whereas *Carnival of the Animals* speaks to all ages, Hungarian Erno von Dohnányi's *Variations on a Nursery Air* for orchestra with piano obbligato is better suited for older children and adults. It is greatly humorous and also satirical. It opens pompously, with a high degree of self-importance, but then the extremely simple "Ah, vous dirai-je, Maman" nursery air is heard. The work proceeds, becoming humorous, then serious, and then back again, as the variations are played out.

Other types of landscapes are of great interest to children. For example, the well-known, thickly forested landscape in which a candy-coated witch's house is located appears in *Hänsel und Gretel* (1893), the operetta by German composer Engelbert Humperdinck (1854–1921). It begins with the simple and familiar "prayer," which is later developed in a livelier manner, with echoes of the witch's music. I once played the part of the witch, and the audience of mostly children cheered when I was pushed into the oven!

THE MAGIC OF BIRDS

Mention of birds was made in the last section. Composers through the centuries have included references to birds singing. No discussion of music and nature would be complete without reference to Ottorino Respighi's *The Birds* (1927) and to Olivier Messiaen's (1908–1992) deep concern for bird songs. Respighi's dance rhythms in *The Birds* are captivating, being based on music by others. His charming orchestrations are quite delightful, representing the dove, the hen, the nightingale, and the delicate cuckoo. Messiaen had the gift of being able to hear birds' songs and then write them down on manuscript paper as music but, intentionally, "one, two, three, or even four octaves lower." As with other composers, he thought in terms of colors. Some of his music was directly related to specific flowers, including "yellow irises" and "purple foxgloves" (see Johnson 1975, p. 101), but I must cut short a fascinatingly complex matter, for at this juncture my focus is on bird songs.

Messiaen studied birds in France, North America, and Japan, and, using gramophone records, in other parts of the world. This love was but one aspect of his intense love of nature in general. His *Chronochromie* (*The*

Clouds of Time) (1960) is an homage to the Alps. He was fascinated by all sounds, rhythms, and colors in nature, but bird sounds fascinated him most. His *Reveil des oiseaux* starts at 3:00 AM and continues well into the day. The dawn chorus of birds contrasts with silence at noon. The work includes references to "all the birds of our forests." In some of his musical portraits, he "literally paints a portrait of the whole bird" (Johnson 1975, p. 119), whereas at other times he merely hinted at what bird he had in mind. His *Turangalîla-Symphonie* (1949) is his best-known work. Without delving into its story, I will explain that the title is a combination of two Sanskrit words that, taken together, mean, depending on one's source, "a song of love," "a hymn of joy," or "the speed of life." For large orchestra, including a large percussion section and an Ondes Martenot, the work has immense rhythmic complexity and inventiveness. The removal from "reality" makes it meaningful to mention Messiaen's use of bird sounds in this amazing symphony in this section, for the soundscapes are very much in the realm of the imaginary. Indeed, some listeners hold that his symphony reaches near mythic expression. It is to myths and legends, or, simply, storied pasts, that we turn to next.

ANCESTRAL LANDSCAPES

Considerable discussion could take place around Wagner's use of myths in his gigantic operas, but opera is not our focus. Not just Europeans have been concerned with the role of myths in their music; hence, I turn now to some important issues pertaining to composers who happen to be indigenous persons. Myths are often embedded within legends, or, to put it another way, within traditional tales that explain histories, beliefs, practices, or natural phenomena.

Indigenous peoples worldwide generally retain close ties to their environments, as already briefly noted. The Maori in New Zealand, for example, have a living concern for legends and myths that have great continuing significance. Many compositions for orchestras by New Zealand's composers make reference to that country's landscapes, including Douglas Lilburn's *Aotearoa Overture* (1940). That and his several symphonies have numerous locational and landscape references, but the link to Maoridom is unclear. In contrast, Maria Grenfell's *Stealing Tutunui* (2000) is explicitly based on a Maori legend. Christopher Blake's *The Coming of Tane Mahuta* (1987), a piano concerto, is based on the Maori creation legends. "Each movement depicts a stage in this mythology—the coming of existence, the coming of life, and the coming of death" (Blake, Centre for New Zealand Music). These works and many others call upon stories of ancestors and their associations with specific parts of the landscape.

Important rituals are associated with death. One such ritual is beautifully embedded in how a particular work came to be written. Gillian Whitehead (1941–), a Maori composer in Dunedin, New Zealand, wrote *Ahotu* (*O Matenga*), a work of "alluring mystic charm" for the unusual combination of flute, trombone, cello, piano, celeste, and percussion, which traces the journey of the soul after death. The work is dedicated to the memory of Whitehead's father. *Ahotu* is one of the phases of the moon, which Whitehead uses as a sort of opus number. She writes (personal communication, June 10, 2001), "in Maori traditional belief, after death the spirit travels northwards through the country; from Cape Reinga [at the extreme northern tip of North Island] the spirit travels northwards to Hawaiki, probably the ancestral homeland." Further, "when someone is dying they ask for a taste of a specific food to sustain them on their spirit journey." When Whitehead's dying father asked her for a particular food, it was not possible for her to get it because it was Sunday and the shops were closed. Instead, she wrote the composition. It must have been a magical moment when she handed him the work, for I am sure he felt great pride in knowing that she was giving him something so heartfelt.

Most Maori composers today are closely involved in the "real Maori tradition," in which the transmission of experience and accumulated lore is "almost exclusively oral." Barry Mitcalfe (1974, p. 1) states that this oral transmission is carried out "through formulaic expression, songs, chants, and rituals—charged particles conveying the deeds, dreams, desires of a people." As Hirini Melbourne observes, referring to days of old, "There wasn't anything music didn't touch" (quoted in Shieff 2002, p. 137). The sacred and ceremonial songs are held "in common by all the separate descent-groups of Maori society. These *Karakia* contain the core of Maori culture" (Mitcalfe 1974, p. 1; see also Shieff 2002, pp. 137–49). Given the importance of oral transmission, it is perhaps understandable that Maoris generally have not embraced Western orchestration. This is not to say that Maoris have neglected Western music, for some Maori composers, including, notably, Whitehead, and various artists have excelled nationally and internationally. One such artist is Dame Kiri Te Kanawa, the opera super star who, interestingly, upon returning recently to New Zealand from an extended stay in Europe, took to recording Maori songs she had learned during her childhood, thereby completing a circle.

Ties to ancestral land are also vitally important across the Tasman Sea in Australia, but Aboriginal musicians to date seem not to have written music for orchestras. However, as in New Zealand, Canada, and the United States some young people are learning Western composition, this situation may soon change. The Aborigine's didgeridoo—the long, hollowed-out tree branch that is blown into—has a distinctive melody characterized by musicologists as a "tumbling strain" (that is, a sound that descends again and again in a deep raspy manner). Played along with rhythm sticks, men singing, and hand clapping, the pulsating sounds are distinctive. (On Aboriginal music, see Marett, Ellis, and Gummow 2001.) The didgeridoo has entered into some Western music, not only in Australia but in the United States as well. As noted in chapter 2, Peter Sculthorpe uses Aboriginal melodies in *Kakadu*.

Little orchestral music exists from indigenous peoples in the United States and Canada. Clifford Crawley's 1978 work called *Tyendinaga* is a rhapsodic elaboration of an Iroquois lullaby. While not a landscape scene as such, perhaps at least we can imagine the work being played on a CD in a baby's bedroom. The nonindigenous American Edward MacDowell (1861–1908), known best today for his piano concertos and several short orchestral works, developed a strong interest in the music of the indigenous peoples in the United States, notably in the Northeast, where he lived. His romantic, five-movement Suite no. 2, *Indian* (1897), includes several indigenous melodies. The movements are "Legend" (Iroquois), "Love Song" (Kiowa), "In War-time" (using a melody that is common to many people of the northern East Coast), "Dirge" (Kiow), and "Village Festival" (Iroquois). The work invites listeners to reflect on the Iroquois and Kiow peoples and where they live.

Welcome to PauWau: A Gathering of Nations (2000) is a work of greater significance. Brent Michael Davids drew upon his European classical training and his heritage as a member of the Mohawk Nation (from the New York State region) and sought to avoid the so-called American sound of Leonard Bernstein, Aaron Copland, and others. The symphony represents a powwow, a traditional forum for meeting people from many tribes. Note the contemporary spelling, "powwow," which differs from the old form used by Davids.

A current-day powwow takes place in an open field or campsite. It normally lasts two or three days. The soundscape is unique, being an amalgam of sounds from daylong, insistent drumming, singing, and dancing by people of all ages. There will be a period of intensity and then a slacking off, with perhaps only one or two drummers keeping a beat; after a break, though, they will be rejoined by other drummers and, likely, also dancers and singers, and another period of intense singing and dancing will begin. In addition, there are craft displays, food, and socializing. People from many Indian nations attend these events, no matter which nation organizes it. Indeed, people from many indigenous nations travel to powwows from far and wide; thus, for example, it is common to find people from Canada's West and even the U.S. Southwest participating in a powwow in eastern Canada. Due to the diverse sources of participants, the landscape is one of tremendous color from the many varied traditional clothes and headgear that people wear. The retelling of myths and traditional stories is at the core of the event. The landscape and the associated soundscape are like nothing else on earth. To attend a powwow is a wonderful, healing experience, due in part to the acceptance and celebration of difference. Powwows are held in many places, all across the United States and Canada each summer. With this as background, let us return to the composition in question.

Davids reflected on the sights, sounds, and symbolic significance of powwows and wrote his music in a manner that evoked what was in his mind. His percussion scoring calls for changes, from sometimes consistent to sporadic playing. He also uses dancers as sound makers; bells are sewn onto their traditional costumes and ring as they dance. Further, a choir sings in a distinctive, nasal manner to represent a traditional Native American vocal sound. *Welcome to PauWau: A Gathering of Nations* surely is challenging to play for any player trained solely in Western classical music, for the complicated, changing rhythms are tricky, as is the tonal nature of the work. The symphony has fifteen movements, each representing something of a powwow's sounds, colors, and magic atmosphere. Not only does this symphony portray a vivid cultural event that creates a distinctive landscape, but it does so using music that is at once basic to indigenous culture, yet links it to ways of music making in the classical sense using classical orchestral instrumentation.

The works just discussed are all grounded in beliefs or stories of the past, some of which have achieved mythic dimensions. The interplay between the mythic past and the present is both immediate and real for those who believe, as the next section will show.

MYTHS AND NATIONALISMS

Myths speak to the links between the past and the present, and usually to links between people and their land. The word "myth" here refers to traditional stories of mystical veneration and significance for particular cultures. A myth is a "sacred" narrative for members of the culture, who accept the myth's validity as an explanation for things in the culture's distant or founding past. Some myths take on epic form, as with some Greek and Scandinavian myths. In many instances, it is difficult to distinguish legends and fairy tales from myths. In general terms, the characters in myths are supernatural beings, ancestors, or heroes who fulfill primordial needs, as Wagner seeks to reveal in his operas. Myths bring the unknown into relation with the known. These thoughts can be tied to the notion of nationalism, a sentiment powerfully placed at a conjunction of a nation's past, present, and hopes for the future. Nationalism has as one of its key elements a territory, held or coveted (Shafer 1955). A territory is real, having bounds and content; yet, it is also a human construct, for the people of a nation have a psychosomatic relationship with their territory (Gottmann 1973). Most nations have or desire a state. Territory is fundamental for the very existence of a state, offering as it does both security and opportunity. The notions that tie people to their particular national location on the earth's surface are "real"—for those who believe (Knight 1982b; Herb and Kaplan 1999). The tie does not exist for others, other than perhaps in a negative sense, as something to oppose. Since a nationalism is very much of the mind, its discussion is included here as something that is imagined.

NATIONALISTIC CELEBRATIONS

Music that speaks to a nationalism can provoke excited responses. Berlioz's *Rakoczy March* (1846), originally written as a *Marche hongroise* (*Hungarian March*) and used in *La damnation de Faust*, created an electrifying reaction when first performed in Budapest under the composer's direction. Berlioz recorded the 1846 event in his memoirs:

> When the day came my throat tightened, as it did in time of great perturbation. First the trumpets gave out the rhythm, then the flutes and clarinets softly outlining the theme, with a pizzicato accompaniment of the strings, the audience remaining calm and judicial. Then, as there came a long crescendo, broken by dull beats of the bass drum, like the sound of a distant canon, a strange restless movement was to be heard among the people, and as the orchestra let itself go in a cataclysm of sweeping fury and thunder, they could contain themselves no longer; their over-charged souls burst with a tremendous explosion of feeling that raised my hair in terror. I lost all hope of making the end audible, and in the encore it was no better; hardly could they contain themselves long enough to hear a portion of the coda.

The *Rakoczy* tune, Berlioz knew, spoke to the powerful patriotic feeling that Hungarians felt for themselves as a distinct people. Another well-known score, *Finlandia*, by Jean Sibelius, received such an uproarious reception among Finns that the Russians, then controlling Finland, banned the work from public performance (see below). Some nationalistic music, when performed "elsewhere," can provoke powerful emotions, even among nonnationals, as happened in 1966 following a performance in Chicago of Bedřich Smetana's Czech epic *Má Vlast* (see pp. 171–72).

The notion of "My Country" above all others is inherent in all nationalisms. A glance at any collection of national anthems provides more than adequate evidence of this (Boyd 2001, vol. 17, pp. 654–55; "National Anthems" 2001, vol. 17, pp. 656–87). Love of country, love of place: this can be expressed in many ways, not least by the written word, as when

Shakespeare has John of Gaunt (in *Richard II*, Act II, scene i) say, "This royal throne of Kings, this scept'red isle, . . ./This other Eden, demi-paradise, . . ./This fortress built by Nature for herself . . ./This precious stone set in the silver sea, . . ./This blessed plot, this earth, this realm, this England." Hubert Parry agreed, as evidenced by his rousing hymn to England, *Jerusalem*, as did Edward Elgar in his lively *Pomp and Circumstance* marches, William Walton in *Crown Imperial* and *Orb and Sceptre*, and Henry Wood in *Fantasia on British Sea Songs*, which concludes with "Rule Britannia!"

Love of country can also be expressed in terms of one's response to particular music. Why else do some people get lumps in their throats when their national anthem is sung? Richard Wagner provides us with one of the more excessive explorations of nationalistic feeling stimulated by hearing some music, in his case a performance of Carl Maria von Weber's opera, *Der Freischütz* (*The Marksman*). Wagner wrote afterwards in a German newspaper, "O my magnificent German fatherland, how must I love thee, now must I gush over thee, if for no other reason than that *Der Freischütz* rose from my soil! . . . How happy he who understands thee, who can believe, feel, dream, delight with thee! How happy I am to be a German" (quoted in Taruskin 2001, p. 693). Wagner believed in German mythology and placed it in his many operas, works that came to symbolize "Germanism" (Shafer 1972, p. 213). However, Sydney Finkelstein notes that Wagner "had to an extraordinary degree the national faculty for duping himself with his own words" (Finkelstein 1960, p. 139). Such may apply to most patriots, of whatever origin. Serge Rachmaninov, for example, enthusiastically expressed his love for his motherland, Russia, in his Symphony no. 3 with music that has a distinct Russian feel to it.

Much has been written on national idioms in music (e.g., Lawson and Stowell 1999; Salzman 1974, pp. 69–90), though that is not the thrust of my discussion. Instead, I am concerned with expressions of nationalism and landscapes in music. But the *Rakoczy March*, *Finlandia*, *Má Vlast*, *Der Freischütz*, and others are all examples of music that is at once highly nationalistic and, yet, also universal, appreciated by nationalists and, with different sentiments, by nonnationals. We turn now to *Má Vlast*. The work as a whole warrants discussion because it draws so clearly upon the landscape and myths of importance to a people, in this case, the Czechs.

NOT IN THE REALM OF ABSOLUTE MUSIC

Bedřich Smetana (1824–1884) was born in Bohemia. He moved to Praha when he was nineteen and was in that city when, as in most of Europe of the time, nationalistic ferment reigned. Smetana joined others at the barricades against the German oppressors; the revolution was short-lived and brutally suppressed. Smetana, in time, became the chief conductor at the National Theatre at Praha, but it was for his compositions that he became most famous. He gave Czech music its typical national color and rhythmic expression. David Dubal (2001, p. 295) quotes Smetana: "My compositions do not belong to the realm of absolute music, where one can get along well enough with musical signs and a metronome." His several operas were in the Bohemian idiom. The first, *The Brandenburgers in Bohemia*, was enjoyed by patriotic Czechs, though its story annoyed the German-oriented critics. His most famous opera, *The Bartered Bride*, a comic opera, concerns life in the country. A much adored work, it includes an amazing furiant, a polka, and a skočná, all well known Czech dances. The last dance is generally known by its title within the opera, namely, *Dance of the Comedians*.

Má Vlast (*My Country* or *My Homeland*), Smetana's major orchestral work, represents aspects of Bohemian life, real and imagined. The work's six symphonic poems can each stand alone—two (*Vltava* and *From Bohemian Woods and Meadows*) are frequently played as separate works in concert programs—but they make most sense when played as a whole, for they relate a story, or, rather, a series of stories that pertain to the Czech nation.

The first symphonic poem, *Vyšehrad*, represents the huge half-legendary rock that overlooks the Vltava river in Praha. The music tells of the awakening of the poet's dreams of Vyšehrad's glorious past. The opening harp chords represent the castle, the Czech nation, and the Princess Libuše's vision of a glorious future for her people. After a festivelike development, the work concludes with the image of a castle destroyed, a silent ruin standing as a reminder of glorious days of old. The second symphonic poem, *Vltava* (or *Moldau*; see pp. 55–56) describes the river, starting at its two source springs and flowing, larger and larger, until it passes Vyšehrad and then continues north and leaves Czech territory. *Šárka* follows, representing a legend in which a maiden seeks revenge because of her lover's infidelity. As Knight Ctirad and his men come near, Šárka bids her warrior maidens to bind her to a tree so that her call in distress will attract those she hates. Ctirad hears her and comes to her rescue. She is released. Šárka gives a potion to Ctirad and his men, who, after a drinking bout, are soon asleep. She calls upon her warrior maidens to come out of hiding, and they slaughter the men. Next, in *From Bohemia's Woods and Meadows*, Smetana describes the beautiful Czech countryside, with its dimly lit woods and bright open fields, and presents a warm idyllic image of the simple joys of life in the countryside, using broad melodies and a lively polka rhythm. *Tábor* then portrays the most famous era of Czech history, the fifteenth-century Hussite wars. The Hussite hymn, "Ye Who Are Warriors of God," provides the theme, which represents the uncompromising resistance Protestant Hussites showed in defending Tábor—and their beliefs—against their numerically superior foes. *Blaník* concludes the cycle, repeating the giant chord with which *Tábor* ends. It develops the same Hussite motif; however, the Hussite men quarrel and fight among themselves, so they retreat to the hollow Mount Blaník to sleep. A pastoral section depicts the mountain as they lie asleep inside, awaiting the time when they will awaken and ride out in joyful triumph to save their fatherland—*Má Vlast*. The Hussite chorale joins toward the conclusion with a joyous rendition of the opening theme of *Vyšehrad* in, as František Bartoš (1951, p. xv) puts it, "the final apotheosis of a resurrected people, and of their future happiness and glory."

Má Vlast was first performed in its entirety in 1882 and dedicated to the city of Praha. The work is often dark in color, reflecting not just Czech legends but also the fierce days of the Hussites; yet, in marked contrast, *Vltava* and *From Bohemia's Woods and Meadows* shine with captivating brightness in their vivid representations of landscapes. Czechs quickly accepted *Má Vlast* as an expression of their nationalism: by presenting a beautiful and colorful representation of the Czech rural landscape and their storied past, the work identifies *their* national heritage and *their* national homeland. One might imagine that the full cycle would speak only to Czechs, but the work has universal appeal, and it appears periodically in orchestral programs around the world.

A FINN FOR FINLAND

Jean Sibelius also lived in a politically volatile time. Some of his writings very clearly were born out of political struggle, for Finns lived at that time under the fist of Russian imperialism. In 1891, Sibelius wrote his *Kullervo Symphony*, his first major work, like others of his works to follow, to draw inspiration from the *Kalevala*, a collation of Finnish legends and folklore. The symphony stood as a testament to a separate Finnish identity in a land in which Russian rule was hardening. Next, he composed *En Saga* (1892), which was distinctly non-Russian in nature. Shortly thereafter, in 1893, Sibelius wrote the first of his openly patriotic works. It originated as part of about a dozen songs and orchestral pieces written as incidental music to accompany a series of students' historical tableaux of Karelian subjects in November 1893. Russia controlled the Karelia region; Finns wanted to control it for themselves. As Guy Rickards (1997, p. 56) notes, the music received considerable acclaim at the premiere in Helsinki, "although the composer believed this was due more to burgeoning national sentiment than musical appreciation." Six days later, in concert form, eight of the works were performed to further acclaim. Sibelius then included the core ideas in his *Karelia Overture* and three-part *Karelia Suite*. The three-part suite, the second part of which is thought to be based on Finnish folk tunes, quickly became one of the composer's most popular works.

Moved by a spirit of Finnish resistance to the Russians, Sibelius then composed two tone poems, and in 1895, he composed *Lemminkäinen* (the *Four Legends* cycle, one of which, *The Swan of Tuonela*, is discussed in chapter 7), in all of which Finnish sound colors are striking. It has been suggested that Sibelius adopted the Russian symphonic model, but he then clearly made it his own with the use of evocative orchestration and programmatic developments that set him quite apart from any Russian influence. As Rickards has observed, his *Kullervo Symphony* and *Lemminkäinen* had distinctly Finnish colors, whereas his Symphony no. 1 (1899) "might even have suggested an artistic rapprochement with the Russians," but it actually "sounded more like a Finn beating them at their own game" (Rickards 1997, p. 66). The major act of resistance came in 1899, following the closure of newspapers due to trouble with censors. One such newspaper held a three-day festival in Helsinki to raise money for the Press Pension Fund. Robert Layton (1983, p. 12) notes that the event was "seemingly harmless" but "laden with political fervour" and that it assumed "the character of a patriotic demonstration." Sibelius wrote music for the festival's concluding series of seven tableaux, which illustrated Finland's past. The opening scene depicted Finnish involvement in the Thirty Years War (of 1618–1648); the other scenes portrayed different aspects of Finnish folklore. Three parts of the incidental music were issued as *Scènes historique* (1899). The final tableaux attracted the most attention. As he had done with the *Karelia Suite*, Sibelius made the concluding music markedly patriotic. Known initially as *Finland Awake*, the final piece was published separately as *Finlandia*, the title having been suggested by the author of a letter to a Helsinki newspaper. This was the work that the Russians banned.

Representative of neither the power of expression nor the depth of emotion that empowers his seven symphonies, *Lemminkäinen*, *Tapiola* (named after Tapio, god of the forest in Finnish mythology), or *En Saga*, *Finlandia* has been dismissed by some as shallow and trivial since it speaks to the raw emotion of nationalism. As such, it is no better, according to some critics, than Walton's and Elgar's compositions for the English and, for the Russians, Pyotr Ilich Tchaikovsky's *1812 Overture* (which sees the French—*La Marseillaise*—defeated by the Russians—*God Preserve the People*). Such critics obviously dislike musical nationalism, for that is what they say *Finlandia* is. Even so, as a statement in music of a people and their land, it spoke to Finns of the day, and it has spoken to countless others ever since, including non-Finns around the globe. While at one level highly nationalistic, at another level, it is universal. So moved were some people that they adopted the tune for a hymn that appears in British and North American hymnaries.

A NORWEGIAN TOUCH

Not all nationalistic music is bold or bombastic. Boyd Shafer (1972, p. 213) observes that Claude Debussy and Ravel, "though not violent nationalists, wanted 'good French music.'" The same, with respect to Norway, can be said of Edvard Grieg (1843–1907). Of Scottish origin (for his great-grandfather had emigrated from Scotland to Norway in 1770), Grieg became Norway's premier national composer. He studied piano in Leipzig, Germany, at the age of fifteen. Later, following on from early tentative steps toward a Norwegian national music by Otto Winter-Hjelm and Rickard Nordraak, Grieg created a national portrait of outstanding quality. Unlike Sibelius, who used folklore as the basis for his music, the Norwegian composer used folk music. When in his midtwenties, Grieg saw a copy of *Aeldreog nyere norske fjeldmelodier (Older and New Norwegian Mountain Melodies*) by Ludvig Lindeman, whereupon he arranged piano versions of some of the melodies, some of which were later orchestrated. The first, *Lyric Pieces*, op. 12 (1867), included "Norsk Melode" and "Folkvise" (Folk Tune), as well as "Faedrelandssang" (National Song) for which a colleague subsequently wrote patriotic verses. In all, he wrote ten sets of lyric pieces, including *Lyric Suite,* op. 43 (1886) for piano, which (orchestrated in 1904) has four movements, including "Norwegian Rustic March" and "March of the Dwarfs." In 1869, he traveled abroad to Italy. While in Rome he met and played some of his music for Liszt, the Hungarian nationalist composer, who was most encouraging. Once home, he wrote two cantatas, with orchestral accompaniment, and he gave still further evidence of an awakening nationalism by helping in 1871 to found the Christiania (Oslo) Musikforening for the promotion of orchestral music. He took to gathering folk tunes from various rural settings, notably in western Norway. Earlier, in 1867, Henrik Ibsen had invited Grieg to write incidental music for his vast, five-act drama, *Peer Gynt*. Ibsen was a fussy man, which was the principal reason Grieg took eight years to compose the music, which was completed in 1875. Twenty-three numbers were used for the stage performance in 1876, from which Grieg selected eight to be included in his *Peer Gynt*, Suites nos. 1 and 2. It is this music more than all else that made Grieg Norway's nationalistic composer.

Peer Gynt, Suite no. 1, opens with "Morning Mood," seemingly based on the music played on the Norwegian hardanger (a traditional fiddle). In the play, the music served to introduce a scene set in southwestern, coastal Morocco. The opening flute line is one of lightness, a feeling that is carried throughout the work, as one senses the sun rising (perhaps reflecting Grieg's immediate reality of the sun appearing over the mountains and flooding a fjord with its gentle, northern light). There follows "Åse's Death," Åse being Peer's mother. The strings give one a sense of ache, of a mournful expression of pain. "Anitra's Dance" sees Peer Gynt, in Arab clothes, being welcomed as a person of consequence, but the opening dance, also played on the strings, is a mixture of brightness and weariness. Finally, with the bassoon leading the way, "In the Hall of the Mountain King" transports us to Norway, to the land of trolls and of the Old Man of the Dovre, whose daughter Peer desires and whose kingdom he covets. Backed by a pulse from the strings, other woodwinds join. The tempo gradually picks up, until at the close, with a great swell on the timpani, the work comes to a dramatic end. Suite no. 2 opens with "The Abduction" (sometimes called "Ingrid's Lament"), for Peer has abducted her from her wedding and is about to betray her. This work opens with a flourish that gives way to a mournful exposition on suffering and loss. "Arabian Dance" follows, with Peer once again garbed in Arab clothing. Flutes to the fore, with percussive interludes, the music weaves a magic spell. In "Peer Gynt's Return" (or "Homecoming"), Peer, now an old man, returns to Norway, his own country. This is the most dramatic of the movements. We sense that he is on a turbulent sea with surges and swells. The movement ends on a short, quiet note, not one of rejoicing. The music gives way to a beautifully rich tune, "Solveig's Song," in Grieg's typically understated manner. It is at once bold and yet simple. Peer is accepted home again, getting just a brief glimpse of the girl he left behind (now a middle-aged woman) before the work ends.

Norway was under close Danish rule until 1814, when control passed to Sweden, which maintained control until 1905. Grieg, thus, also lived in a land dominated by others. In contrast to Sibelius and Smetana, he chose to give a quiet voice to his nationalism. He was strongly influenced by folk tunes, as noted, and he was also profoundly influenced by his time in western rural Norway, not least at the fjord at Lofthus, where he would wend his way on foot along the shore to a knoll on which he kept a piano in a one-room, wooden cabin. It was there that he composed much of his music. The setting was exquisite, with the dark, deep fjord in the foreground and the Folgefonna glacier behind it. The sun, at different times of year and day, would shine brightly or cast vast shadows. Aimer Grønvold states that Grieg developed "an intense and indissoluble relationship between the environment he lived in and the music he created" (quoted in Herresthal 2004, p. 2). Grieg composed what he saw and experienced in an intimate manner that speaks at once to Norwegians, calling as it does on their folk tunes, but also to others. His music transcends time and space in a special way. His many works are all mostly miniatures, finely defined works of short duration. Taken singly or together, they gave voice to a nationalism that at once challenged the imperial presence of the Swedes but did not threaten them. Such is another type of nationalism in music, one that contrasts markedly with *Finlandia* and *Má Vlast*. However, Grieg's music, like that of Sibelius and Smetana, is warm and melodic, rich in lyrical beauty and originality, and, especially like that of Smetana, characterized by delightfully ingenious modulations and rhythms. Each composer's work, however, spoke to different national traditions.

OTHER NATIONAL REGIONS

A huge literature of music pertains to national identities, not all of which can be cited and discussed here. A few additional examples of nationalist compositions include Georges Enesco's lively *Rumanian Rhapsodies*, Liszt's *Hungarian Rhapsodies* (the second of which incorporates Roma, or Gypsy, influences), Antonín Dvořák's *Slavonic Dances*, Frédéric Chopin's *Polonaises* (a polonaise is a Polish national dance) and *Mazurkas* (another Polish dance), and Elgar's *Pomp and Circumstance Marches*. There are also Russian works, including Mikhail Ivanovich Glinka's *A Life for the Tsar* (with its concluding jubilant, marchlike nationalistic anthem), Borodin's *Polovtsian Dances*, Tchaikovsky's *Marche slav* (which includes *God Save the Czar*) and *1812 Overture*, and Mikhail Mikhaylovich Ippolitive-Ivanoff's *Caucasian Sketches*. The last of these many works is written in four parts: "In the Mountain Pass," "In the Village," "In the Mosque," and "March of the Sirdar." "In the Mountain Pass" is especially geographic, for one can well imagine the movement of people as the music is performed. *Caucasian Sketches* was one outcome of Ippolitive-Ivanoff's studies of Georgian music while he was a conductor of the symphony orchestra in Tiflis (Tiblisi), Georgia. That an outsider can write nationalistic music should not surprise, for as with Berlioz's *Rakoczy March* (mentioned earlier) with respect to Hungarians, the blood of Spaniards—and others—is stirred by Russian Rimsky-Korsakov's *Capriccio espagñol*, Frenchmen Debussy's *Ibéria*, Ravel's *Rapsodie espagñole*, and Emmanuel Chabrier's *España*. It took a while for Spanish composers to develop their own voice, but in time they did, as evidenced by Isaac Albéniz and his *Suite espagñola* (Andalusian influences) and *España*, Joaquín Rodrigo and his *Concierto de aranjues*, Enrique Granados and his *Andaluza* (in honor of Andalusia), and, most especially, Manuel de Falla and his *Ibéria*. And so I could continue. . . .

LOCAL VERSUS UNIVERSAL

There is a tension between the local and the universal. Cedric Thorpe Davie felt that one had to compose for the local; universal acceptance would follow—if the work was good. So it has been for "nationalist" composers (see, for instance, Engel 1866; Vaughan Williams 1934, 1996) who have composed what might be called "localisms," which are now known far beyond the territorial limits of the lands and nations for whom the works were first intended. Richard Taruskin (2001, vol. 17, p. 692) suggests that the "discovery of the folk" was critical in the development of national music. There was, he states, a "recycling of an ancient idea, that of primitivism," whereby qualities of past cultures were deemed to be worthy of contemporary (re)development. In this, as in the previous chapter, comment was made on the use of folk tunes. Such tunes, especially when tied to myths and other stories of the past, prove powerfully evocative, stimulating people's attachments to the group, the nation, and their nationalism, set as they all are within particular territories. This chapter has shown that many imagined places have found expression in music, music for the ages, music that is at once local and universal. Some such music pertains to the search for meaning in difficult environments, and it is to that topic that we turn next.

6

Searching for Meaning in Landscapes of Extremes

Landscapes of extreme can be dangerous yet attractive: dangerous because of the threat to human occupancy and, perhaps, to life itself; attractive because of a perceived pull to the feature or place, possibly due to religious or nationalistic beliefs. The word "extreme" implies landscapes and events in nature that in some way are special or problematic for humans, as well as certain places inhabited by creatures of the spirit world. This chapter stresses situations out of the ordinary, including a deep "scar" in the surface of the earth, a descent into hell and satanic cavorting, mountains as sacred places and as sources of inspiration, volcanic activities, a mountain worth climbing, storms on land, and men who challenged nature only to lose to its mighty power. In addition, the issue of understanding meaning in landscape is discussed, for all through history, people have sought to rationalize their existence, as part of normal living and, especially, when faced with extreme situations. Songs of praise are not considered, but it may be useful to remark that even Christians refer to climatic and geological stressors, such as famous English hymn writer and composer Charles Hubert Parry's call upon God to "speak through the earthquake, wind, and fire." Any further comment of a theological nature will apply only to the music discussed.

THE GRAND SCAR

The Dead Sea may be the deepest portion of the earth's surface below sea level, but the most spectacular undoubtedly is the Grand Canyon, a huge scar that cuts across the landscape of the southwestern United States. The physical size and geomorphological diversity of the Grand Canyon is breathtaking. Whether to be photographed, painted, expressed in poetry and novels, described and analyzed by scientists, or represented in music, the magnificence of the Grand Canyon continues to be both a magnet and an illusive mystery. The character of the canyon constantly changes as the sun and clouds move, with colors and topographical shapes seeming always to become something other than what they just were. The best-known attempt to capture this illusiveness in musical form is by Ferde Grofé (1892–1972).

Drawing upon many years of first-hand involvement with the Arizona landscape, including its people, flora, and fauna—and also its bars (saloons), where he played piano for a living—Grofé wrote a five-movement tone poem entitled *Grand Canyon Suite* (1931) for full orchestra. He was struck by the richness of the land, the optimism of the people he came into contact with, and the meaning all this had taken on for him. When he realized that he had developed a spiritual attachment to the place and it inhabitants, he decided to place his observations and understanding in music.

The *Grand Canyon Suite*'s first movement, "Sunrise," begins quietly in the gloom of darkness, and then, with the gradual appearance of light as the sun rises, the orchestral sound grows in a long crescendo until, reflected in a wonderful burst of sound, the light reveals the immensity and glory of the canyon. In "Painted Desert," Grofé portrays this region as filled with amazing landforms and a wonderful array of pastel shades of red, yellow, and gray, whose hues change constantly as the movement and intensity of the sun alters through the day. As we listen, we can almost see the landscape, so clever is Grofé's use of shifting orchestral colors. "On the Trail" centers on the vivid caricature of a sure-footed, but stubborn, mule given to a periodic "hee-haw" as it takes a person down the seven miles (from top to bottom) of twisting trail. The clip-clopping and occasional trot are clear, as is the sudden stop and loud "bray" when the stubborn beast declines to continue. Encouragement is given, and the ride resumes on down the narrow path, with deep vertical drops to the canyon floor below just inches away. Missing from this movement is a sense of what it means to be on the trail, other than in the sense of effort spent—and the passenger's desire for the donkey ride to end. The richness of the changing colors on the walls that are so intimately close as you go up or down the trail is not present. It is easy to imagine Claude Debussy creating a totally different image, but in his concern for the subtle changes of color, he likely would have omitted the humor and delightful, awkward sense of movement created for us by Grofé. Grofé portrays the highlight of any day at the Grand Canyon in "Sunset." A photographer's paradise at any time, the canyon at evening on a clear night surely is heaven sent. From the lookout at the top of the canyon, one sees the bottom getting dark, followed by the lengthening shadows creeping up the craggy walls. Those parts of the canyon walls still in the sun take on a sharpness that is stunning, so complete is the contrast with the darkened areas and so clear is the air. Grofé seeks to convey his sense of delight and amazement by presenting a majestic theme that gradually fades to a serene conclusion as darkness descends. In the final section, "Cloudburst," the listener is taken from imagining a brilliant display of stars in the dark sky above to being soaked as a sudden storm whips viciously across the canyon with shrieking winds, pelting rain, and violent thunder (created using a thunder sheet). Like all storms, this one finally subsides and ends, and there is a return to the peaceful still of night, with the crystal-clear stars and moon above.

Grofé's music has no concern for myth or stories; it is simply wonderfully descriptive of the grand sweep of the huge physical environment, neatly balanced against the detail of a person riding a mule. There is a cleanliness and, yet, also a richness to the sound, which, to my mind, truly reflects the quality of the place. As one hears the music, it is clear at once that Grofé has created a musical picture of a southwestern New World setting, so sharp is his construction of the landscape, devoid of moisture in the air (other than during the short, sharp thunderstorm), which enables one to see the landscape for miles during the day and the brightly shining stars at night. While his sound is rich, there is no lushness of sound from the violins, which would suggest a muggy British or northern European day. Instead, the dryness of the Grand Canyon's air is captured by the highly effective style of playing.

For some, the Grand Canyon is a place of the spirits. For others, it is simply a wondrous place to visit. For a few, it is just a big hole. Meanings attached to or gained from places can be quite personal, even when a group maintains the existence of that place's spirituality. In the event that we assume the descent to great depths can be roughly equal to being dropped into hell—and some tourists in the depths of the Grand Canyon would likely agree because from the bottom, as they look up, they realize they have to go up the long, thin, winding trail to get to the top again—then we can ask what hell might be like. Let us consider a possible answer to such a query.

DESCENT INTO HELL

Painters have sought to convey their understanding of hell and so have some composers. Some of the works on war discussed in chapter 7 surely are expressions of hell on earth. Franz Liszt's Mephistopheles in the 1861 *A Faust Symphony* (after Goethe's poem) is of course the devil. He is the focus of the third movement for which Liszt's printed directions to the players include *ironico*, and, indeed, the nature of the music leads itself to ironic treatment. Mephistopheles jeers, sneers, mocks, shrills, and laughs at Faust and Marguerite, the other two characters. The image of the taunting Mephistopheles is not pleasant; nor is the moment when Faust sells his soul to the devil in return for youth and love. Liszt also wrote *Dante* (1856), with two movements, "Inferno" and "Purgatory," and a quiet, concluding section for women's voices on the text of the Magnificat, "Vision of Paradise." "Inferno" opens with trombones and tuba conveying the musical image of the gates of hell as they illustrate the words written in the score: "Through me you pass into the city of woe. Through me you pass into eternal pain." The horns and trumpets, mostly on one note, add, "Abandon hope, all ye who enter here." Demons can be heard laughing. "Purgatory" opens with twenty-seven measures on one chord. Liszt's *Mephisto Waltz* (1881) and, to a lesser degree, his *Mephisto Polka* (1883) are popular "pops" concert items because of their catchy image of the devil dancing in glee. Similarly, the image of the devil dancing seems to have been in Jacques Offenbach's mind when in 1874 he composed *Orphée aux enfers* (*Orpheus in the Underworld*), for in this two-act work—and subsequent orchestral music—he gave the world the can-can, which typified Offenbach's irreverence and joy of life.

Charles-François Gounod, Richard Wagner, and Hector Berlioz also wrote works based on the legend of Faust. Of these, Berlioz likely could have created the most vivid of any work on Faust, but his *The Damnation of Faust* (1846) is for the most part unremarkable (i.e., when compared to *Symphonie fantastique*). The most evocative sections of his *Damnation* are sometimes played by themselves, including *Ride to the Abyss* and the *Dance of the Sylphs*, in which fantastic creatures, elves, and sylphs charm Faust into sleep so they can fill his mind with voluptuous images of Marguerite. The other work of note, Wagner's *A Faust Overture* (1840), is great for its early establishment of Faust's gloominess (deep brasses and woodwinds, with timpani), followed by a quick but restless melodic line (violins), before the end, when the music seems to "suggest a recollection of youthful love" (O'Connell 1941, p. 595).

César Franck's *Le chausseur Maudet* (1882) relates a story of the depraved and brutal Count of the Rhine, who, for hunting on the Sabbath and many other sins, is condemned to hunt forever in the flames and terrors of hell. Church bells, the count's hunting horn, a chase, a church hymn, and the awesome moment when the count is changed from the hunter on earth to the hunted in hell are vividly portrayed by the orchestra. A vicious hell on earth is described in Mily Alexeyevich Balakirev's gruesome symphonic poem *Tamara* (1882) in which the lustful, amoral, and strikingly beautiful Queen Tamara, in a tower high above the Terek river, uses her siren voice to lure unsuspecting travelers to her side. With exciting rhythms and pacing, the orchestra portrays debauchery and death and, in the morning, the image of a young man's body floating off down the river.

SATANIC CAVORTING

Another Russian's work qualifies for inclusion here since his gathering is devilishly wicked, though perhaps it represents nothing more than recreation time for those from, in this creation, the underworld. Modest Mussorgsky's diabolical *Night on Bald Mountain* (or, as it is sometimes called, *St. John's Night on the Bare Mountain*) is at once barbaric, obscene, ghoulish, dark in texture and meaning, and, dare it be said, thrilling to listen to. Mussorgsky's style is at once raw, wonderfully inventive, and highly original. *Night on Bald Mountain* was orchestrated after his death by his friend, Nikolay Andreyevich Rimsky-Korsakov. The demons in the work may well have been his own, for Mussorgsky suffered dreadfully from alcoholism.

Bald Mountain is Mount Triflag near Kiev, where every June 23, St. John's Eve, there is a celebration of evil presided over by Chernobog, the dark god who appears in the form of a black goat. The fantasy opens fiercely with the strings growling in a highly agitated fashion to suggest subterranean sounds as the bassoons, trombones, and tuba dig deep to give us a sneering, satanic theme. The spirits of darkness then glorify Chernobog and a black mass is performed. At the height of the orgiastic reveling that follows, a bell on the small church steeple in the distance can be heard tolling the hour of six in the morning, so the frenzy has to end, and the spirits quickly disperse. The mountain is returned to normal as daylight arrives (i.e., until the next St. John's Eve). The music marvelously and grotesquely tells us of the ghoulish cavorting and the nakedness of evil. Rather than remain in the realm of the devil, let us now climb out of deep canyons and the subterranean world and consider a landscape of witches.

Not all that is connected to the underworld lacks humor, as British audiences discovered when they first heard Malcolm Arnold's (1921–) overture *Tam O'Shanter* (1955). This work relates in music the Robert Burns poem "Tam O'Shanter." The story is well-known in Scotland, but perhaps not elsewhere. Making his way home on horseback after a heavy drinking session (an inebriated pair of bassoons), Tam encounters a coven of witches dancing at the kirk (church) in the village of Alloway. With drunken bravado, Tam unwisely shows off and, thus, attracts the attention of the witches. At first, they seem to ignore him, but then they give chase. He is not so drunk as to forget that witches, as with all evil spirits, will lose all of their evil spirits if they cross flowing water. He thus heads his grey mare, Meg, for the river. The chase includes the witches flying around (an orchestral, Scottish, bagpipelike dirge), while Tam and his horse ride hard. The chase becomes desperate for Tom when the fastest witch catches up with them. Tam and Meg start across the bridge, and all the witch can do is catch hold of Meg's tail. With one final surge, Meg gets them to safety, but at the cost of her tail, which is left hanging in the witch's hand. With a huge sigh, Tam and horse slow down and stop for a needed rest. Arnold's music is a delight to listen to.

Tam O'Shanter's night of drinking ends on a note of considerable relief. In contrast, after a night spent enjoying a lot of vodka, surely also a witch's brew, a Russian sailor dances (arrhythmic syncopations suggest his drunkenness and unsteady gait) until he is finally able to get it all together, and the work roars to a terrific climax. The work, *Yablonko, Dance of the Drunken Sailor*, or, simply, *Russian Sailor's Dance*, is by Reinhold Glière (1875–1956). Dancing must be a good way to get rid of evil spirits. Manuel de Falla of Spain, like Glière, obviously thought so. De Falla extracted *Danza ritual del fuego* (*Ritual Fire Dance to Exorcise Evil Spirits*) (1915) from his choreographic story *El amor brujo* (based on an Andalusian folk tale). *Ritual Fire Dance*'s conclusion is wild in its sense of abandon.

MOUNTAINS AS SACRED PLACES

Mountains are special places. They can be awe inspiring. They can be the embodiment of sacredness (Bernbaum 1997). Mountains are the points on earth closest to the heavens—and in some cosmologies they are seen to be "the centers" of the earth—so they take on important symbolism. For example, Mount Kailas, the source of the Indus, Ganges, Sutlej, and Brahmaputra rivers, is Tibet's most sacred mountain; Mount Fuji is Japan's most sacred peak for Shintoism and an important location for Buddhism; Greece's Mount Olympus, according to Greek mythology, served as home to Zeus, king of the gods; Moses received the Ten Commandments at the summit of Mount Sinai (or Zion); India's Mount Meru is the center of the Hindu cosmos; in East Africa the Chagga people regard Mount Kilimanjaro as the home of the god Ruwa; Mount Taranaki (called Mount Egmont by European settlers) in New Zealand is sacred to the Maori people; and Mount Shasta in the United States is accepted by the Shasta people as having been created by the Great Spirit. The list is long (see Bernbaum 1997; Brockman 1997), but these few words hopefully are enough to identify the importance of mountains for humans. Following comments about a few minor works, four musical mountains of consequence are discussed, as is some volcanic action.

INSPIRATION AND TERRAIN

Mountains have held special meaning for many composers. For example, after Claude Champagne (1891–1965) saw the Canadian Rockies, he wrote *Altitudes* (1959), a symphonic poem for chorus and orchestra. *Altitudes* is written in a musical style reputedly akin to the painting style of Canada's "Group of Seven" (Archer 1968). The lean sound of the singers as they convey a sense of detail and wonder contrasts stunningly with the powerful sweep of the Rockies as conveyed by the orchestra. Champagne's work begins quietly (in "époque primitive") before moving, via a linking passage ("méditation"), to a strong sense of power and wonder ("époque moderne"). That Champagne appreciated mountains for their special sacred qualities is implied when the chorus, accompanied by the strings, sings both a Huron Indian prayer and a St. Francis of Assisi hymn of praise for the blessing of nature (David Parsons, personal communication).

Some composers have been inspired to write about mountains indirectly, including Canadian Alun Hoddinott (1929–), who wrote *Landscapes* (1975) based on a poem about Snowdonia in Wales, in the United Kingdom. And Harry Freedman (1932–2000), in *Images* (1959), composed musical impressions of three Canadian paintings, including *Blue Mountain* by Lauren Harris. Freedman's composition on *Blue Mountain* opens with a remarkable six-bar crescendo from ***ppp*** (exceptionally quiet) to ***fff*** (extremely loud), presumably to shock us as we turn a corner and there, in all of its glory, stands the mountain. Thereafter, Freedman has an expansive melody played by the violins, which both captures the sharp, craggy edges and the sense of loneliness to be seen in Harris's painting. A keen ear may hear an allusion to the dramatic cloud over the mountain (Proctor 1980, p. 68). American George Strong (1856–1948) wrote a symphony entitled *In the Mountains* (1886), while his compatriot Carl Ruggles (1876–1971) wrote *Men and Mountains* (1924). Of greater consequence, the Swedish composer Franz Berwald (1796–1868) in 1842 wrote *Erinnerung an die norwegische Alpen* (*Memories of the Norwegian Alps*), using folk melodies to convey a sense of traveling to Österland.

MOUNTAINS AND THE MYSTERY OF FAITH

The American composer Alan Hovhaness (1911–2000), as part of a personal search regarding his religious beliefs, constantly struggled with how best to express in music his belief that mountains are depositories of meaning pertaining to the mystery of faith. His search led him to consider many mountains—and other geographical features—in his work. His distinctive voice within the style of music he developed has been criticized as semimystical "Eastern" mimicry combined with reflections of Orthodox and Western church music. Certainly, it is true that Hovhaness used "melodies and imagery of the Orient with counterpoint and formal disciplines of the Occident" to create many musical landscapes (Kastendieck 1964, p. 1008). The resulting landscapes in music are liked or not; there seems to be no middle ground. As part of his personal search for meaning, mountains are an ever-present component of his work.

In his Symphony no. 2, *Mysterious Mountain* (1955), Hovhaness takes the listener on a transcendent journey of the mind. At times hymnlike and lyrical, notably at the start and near the end, the work gradually unfolds over three tied movements in a slow tempo and with increasing strength, eventually to convey a full sense of the majestic grandeur of the mountain. Along the way, we are given details in vibrant color that are at once ephemeral and fragmented, and yet "real" and powerful in their presence. Could this mysterious mountain be Mount Shasta (this is unlikely since Hovhaness wrote *Hymn to Shasta* [1938]) or perhaps Mount Kailus, each of which is at once a real mass of rock thrust up in majestic ways, yet also has mythic qualities attributed to it? Or is Hovhaness exploring a phantom peak? I like to think he is exploring a wondrous mythic place. *Mysterious Mountain* speaks to the soul in an interesting manner, perhaps because of its almost hypnotic style. Of the work, Hovhaness wrote that "mountains are symbols, like pyramids, of man's attempt to know God. Mountains are the symbolic meeting places between the mundane and spiritual worlds" (quoted in Stannard 1999, p. 3). I wonder, does a geomorphologist studying a mountain for, let us say, its landslide activity let such factors enter his or her mind?

It is tempting to think that Hovhaness's work may have inspired Champagne to write *Altitudes* in 1959, for *Mysterious Mountain* was composed four years earlier. Similarities are there, but Champagne's orchestration is tighter and more complex, and their autographs differ. In effect, we are left with an open question. Champagne's mountain is "out west" somewhere in the Canadian Rockies, where it sits as a wonderful example of tectonic uplift. We are not sure where Hovhaness's *Mysterious Mountain* actually is.

VOLCANIC EXPLOSIONS

There is no escaping the location of another of Hovhaness's mountains. It stands in the Cascade Range, within the bounds of Washington State.

Mount St. Helens is usually quiet, but it has the capacity to become a deadly force that can cause wide-ranging destruction, as evidenced by its violent eruption in May 1980. In the high, volcanic Pacific islands of old, "the Gods" could be angry. It is easy to imagine why this is so when one hears Hovhaness's *Mount Saint Helens*, Symphony no. 50 (1982).

The work opens peacefully, in a sense traveling around the mountain in order to give the listener a sense of the grandeur of the rugged, mountainous landscape. Birds sing (flutes), winds blow gently (strings), and trees swing (percussion and woodwinds). Occasionally, however, also quietly, rumbles from the timpani and percussion hint at something going on underneath the earth. The short second movement (mostly strings and the flute) describes Spirit Lake, but this description is rudely interrupted by the increasing power of brasses and timpani, joined by high strings and the periodic striking of a gong. A rattled thundersheet joins the brass in a weird but very effective "blowout" as the mountain erupts. This sets off a long, harshly methodical, timpani and bass drum part, punctuated by gongs, cymbals, and trumpets, at times joined by others, including the strings, beating out the same pattern. One shivers as the magma comes down the mountain, closer and closer. There is a sense of inevitability. This is raw nature, and nothing humans might think of doing can in any way match the power we are witnessing in sound. Finally, however, the destructive power of the volcano subsides. A mournful hymn (horns and trumpets dominant, backed by strings) sings of the wonder of volcanic power and the ravaged landscape, followed by a bouncy, energy-filled rejoicing in the region's grandeur and, in Hovhaness's words, "the living earth . . . the life-giving power that builds mountains, rising majestically, piercing the clouds of heaven" (quoted in Stannard 1995, p. 3). Hovhaness thus ties the uprooting notion of the "living earth" to the "life-giving power that builds mountains," but, to repeat, he also represents calm nature, with birds singing, winds blowing, and trees swaying. With this work in mind, it is relatively easy to understand why many people in the world view violent mountains, such as Mount St. Helens, as some kind of embodiment of a god or spirit.

Dancing on a Volcano by New Zealander Lyell Cresswell (1944–) considers the tension people live with when they are located on a fault line. Cresswell observed that Comte de Salvandy, just before the revolution of 1830 in Paris, said, "Nous dansons sur un volcano" (we are dancing on a volcano). Using that image, Cresswell ties the ever-present threat of volcanic activity along one of the major fault lines in New Zealand with the fault line between emotion and intellect. *Dancing on a Volcano* opens with loud orchestral chords that offer a musical signpost for the work. Six such chords lead to a quiet, reflective section (principally oboe, flute and strings, and bassoon and strings) in which Cresswell reveals the melodic core. A scherzolike section follows (strings, harp, trumpet, and various percussive sounds), which gets louder and is awkwardly dancelike. A crescendo and a diminuendo, followed by another crescendo, lead to further loud orchestral chords that demand notice over strident strings. A "volcanic dance" follows in which, with piano added, the elements are brought together in a vigorous manner. Threatening tuba, basses, and timpani are set against brighter woodwinds and strings, with irregular percussive outbursts. In dealing with the threat of a full eruption and the everyday fact of living with that threat, Cresswell says he is seeking "a harmony of fantasy and logic" (Cresswell 2001, p. 9).

Not mountainous but also volcanic in nature, *Geysir* (1961) by Jón Leifs (1899–1968) presents a powerful, episodic exploration of a geyser. Mention has already been made of Gillian Whitehead's geyser (p. 44). This one is more literal. Leifs, an Icelandic composer, opens with the deep rumble of the contrabassoon. Other lower-register instruments (trombones, tuba, double basses) play quietly. The image is of slowly moving, subterranean water. Higher strings and woodwinds join in as, little by little, the pressure builds until, suddenly, there is a series of powerful eruptions with boiling water shooting upwards, small amounts at first and then larger ones, as the geyser spouts off. One senses limitless power coming from beneath the earth. Some remarkable sounds reveal the spouting of the geyser until the force is spent, and the scene returns more or less to the way it was at the outset. Hjálmar Ragnarsson (1996, p. 7) says the music "portrays the insignificance of man as he stands, weak and helpless, before the troll-like power of the living earth." Like *Mount Saint Helens, Geysir* reveals that raw nature is powerful.

Before turning to a mountain of quite a different character, let us consider the following. The task: take a contour map, mark a line for a cross-section, and extrapolate the varying elevations on a graph. Then, using a piano or other instrument, "play" the contour lines, so that as the elevation increases the pitch goes up, and as the elevation decreases, the pitch goes down. A question for you at the end of the task: is the resulting sound music or mere sound? I will leave that teasing question hanging and, instead, observe that this is clearly not how the next composer approached the task of writing music, although his task resembled, to a degree, the problem just posed in that the orchestra and the audience climb up and down a mountain, the latter being located somewhere in the European Alps.

"UP THE HILL AND DOWN AGAIN"

The second major mountain we will consider here does indeed involve a climb. In fact, just before a performance, a weary orchestral colleague once said to me, "Here we go again. Up the hill and down again with Strauss." This flippant remark does not do justice to Richard Strauss's (1864–1949) geographically and emotionally expressive work. The *Alpensinfonie* (*Alpine Symphony*) (1915) calls for a very large orchestra—as many as 150 players (though most orchestras make do with fewer numbers) including, onstage, the full orchestra, with almost all instruments augmented, four tenor tubas, two harps, and an organ. In addition, there are horns, trumpets, and trombones offstage. A large percussion section is required for the storm scene and for other special effects: a wind machine, a thunder machine, cymbals, a triangle, cowbells, a glockenspiel, a tam-tam, and various drums.

Strauss was able to draw upon the music of others as he contemplated writing his symphony, though it is not known if he in fact knew of, or knew well, the works by such composers as the early Swiss romantic composer Joachim Raff (1822–1882), who wrote eleven symphonies, nine of which were descriptive and one of which was entitled *Alpensinfonie* (1877). And then there was Antonín Dvořák's student Vitézslav Novák (1870–1949), who wrote *In the Tatras* (1902), and César Franck (1822–1890), who wrote *Ce qu'on entend sur la montagne* (1848) for full orchestra (with the violins divided into six parts). Shortly thereafter, in 1856, Liszt wrote what he liked to call his symphonic poem (*Symphonische Dichtung*—a term first used by Louis Spohr [1784–1859]), entitled *Bergsymphonie* (*Mountain Symphony*). As Leslie Orrey (1975, p. 79) puts it, "Liszt climbed mountains metaphorically rather than physically." In contrast, Strauss drew upon his own experiences from many expeditions into the mountains, starting in his youth, as he created in music his fascinating mountain landscape, which carries the audience almost physically up the slope.

Strauss's *Alpensinfonie* is not a travelogue but a traverse during which the listener is opened to geographical diversity and the wonders of nature. The work has both subtlety and power. Unlike the search for God by Hovhaness, Strauss revels in his wonder of the mountain itself and in his amazement at nature's complexity. Listening to the twenty-two continuous sections, which progress from the dark quiet of "Night" to "Sunrise," through the day to "Sunset," and back to stillness and dark in "Night," the listener joins the party, ascending and descending the mountain. Strauss identified his story line by giving each section a title.

The leitmotiv for the mountain is provided by the trombones and tuba performing long chords. Along the way, the climbing party enters a forest (murmuring strings, with horns and trombones) and, in the distance, hears a hunt (horns), wanders up a stream, and comes to a waterfall (full orchestra, with wonderful cascades of sound) in which is revealed a rainbow and an alpine sprite. Thereafter, the party walks through lovely, flower-covered meadows (a *pastoral*), reaches high pastures where cows spend the summer, gets lost in thicket and undergrowth (which happened to Strauss while climbing when he was a boy), carefully crosses a glacier, endures some dangerous moments, and then arrives at the summit, which is announced by the trombones in a triumphant celebration of the awesome sense of height and wide-open space. The moments of reflection about the mountain and the astonishing view give way to a hesitant oboe, which signals the frailty of humans amid the vastness of nature. The woodwinds warn of a coming storm, as the strings suggest the arrival of the mist, which obscures the sun. A terrible storm occurs as the climbers frantically descend by a shorter route than the one taken to climb the mountain. The storm is one of the biggest in music, filled as it is with amazingly intense flashes of lightening, thunder that crashes and rolls, and torrential rain blown by high winds sweeping around the mountain. Once at the lower reaches, with the storm dissipated, the party witnesses a lovely sunset and rests upon the return to the peace of night (theme repeated from the opening). It has been a long day, though the music takes only about fifty minutes! With the exception of a reference to yodeling in the meadow scene and to the hunting horns, the only implied human presence is the climbing party, though none of its members is identified in any way. Apart from the challenge of the tough trip upslope and down again by the climbers, the mountain is revealed by its physical form, the geographical variation of physical features (including the meadow with flowers, the forest, a high pasture, rugged rocks, streams, and the waterfall), and physical processes (wind, water, and ice erosion). The work, in effect, is a wonderful case study in physical geography.

As noted, little human-environment interaction occurs in this symphony, though the references to getting lost and the dangers of the glacier serve to remind us that high mountains can be dangerous. *Echelles de glace* (*Ice Ladders*) by New Zealand composer Christopher Blake is also a poignant reminder of this reality. Written in 1992 in memory of a musician who died in a climbing accident on the Matterhorn, this elegy for large orchestra is based on rising scales that represent the ice ladders of the work's title.

Strauss made no direct mention of any spiritual quality associated with his mountain, other than his obvious sense of awe at nature's diversity and awesome power. Perhaps the mountain, the traverse, and the setting sun are all metaphors for life. Strauss was a boldly self-confident man, so it is interesting to reflect on what this mountain meant to him, emotionally if not also philosophically. Some dismiss the presence of any spiritual quality within the work by suggesting that it simply is a reflection of Strauss's admiration for Gustav Mahler, who died while Strauss was composing the work. I sense there is more to it than just that. Indeed, after I had written the above, I read Michael Kennedy's excellent guide to the

Strauss tone poems in which he concludes that Strauss recognized that an epoch of orchestral writing was closing, and since the work essentially omits "the human factor, humorous, tragic, fantastic or erotic," it represents an epitaph to what had been an important style for more than a century (1984, p. 61). Further, Kennedy feels that Strauss's comment that the *Alpine Symphony* expresses the "worship of nature, eternal and magnificent" was the "nearest he ever came, or ever would have come, to a religious declaration" (1984, p. 61). With this interesting comment about reverence for nature in mind, let us turn to considering storms.

STORMS ON LAND

Some storms at sea were considered in chapter 3, and I have mentioned in this chapter the storms by Grofé, whose storm is short, wild, and sharp, and by Strauss, in *Alpine Symphony*, which is perhaps the most violent storm (and certainly the loudest) in music. Numerous other composers have been attracted to the challenge of creating storms in their music. You may well have your favorite. I have several favorites. Of those over land, there are five: those by Grofé and Strauss and three by two other composers, to whose works we now turn.

With respect to Ludwig van Beethoven's *Pastoral Symphony*, Symphony no. 6, I observed in chapter 2 that he was inspired by the rural landscapes around him. The third movement of his symphony includes one of the most stunning storms in all of musical literature. Unlike Strauss, who used a huge orchestra, Beethoven used the (then) standard (chamber) orchestra: strings plus two flutes, two oboes, two clarinets, two bassoons, two horns, and two trumpets, though, for the third movement only, he added piccolo, bassoon, three trombones, and timpani. That was it. With these forces, Beethoven created a remarkable soundscape.

The movement is surprisingly short; just 155 bars of music, in 4/4 time. In the second movement, people had been laughing and dancing, but in the opening scene of the third movement, we hear two bars of the dull rumbling of the double bass followed in like manner by the second violins (both ***pp***), with first violins calling out short warnings of impending danger. The message that a storm is coming sends people scurrying for cover. An artfully sculptured transition from quiet to an eruption of sound occurs when, with a huge crescendo by the strings in bars nineteen and twenty, a full-orchestral chord (***ff***) announces the first blast of the storm, which soon becomes a tempest. We can almost see, and we can certainly sense, the pitch-black clouds scurrying across the sky, rain pouring down, winds blasting fiercely, and trees swaying violently in the immense gusts, with sharp convulsions of thunder. The sound picture of thunder (timpani) and winds (strings and woodwinds, topped by the piccolo) is astonishingly vivid. The innovation of having the piccolo shrieking as if it were the high wind battering the tops of trees is remarkably effective. Hector Berlioz ([1862] 2000, p. 75) felt that Beethoven's "awful cataclysm, the universal deluge—the end of the world . . . literally produces giddiness; and many people, when they hear this storm, can scarcely tell whether their emotion is one of pleasure or of pain." Then, rapidly, this short, sharp, yet complete, storm ends, and without a break, the movement quietly flows into the symphony's fourth and final movement (played without any of the just-used "extra" instruments). While the storm created by Strauss is stunning in its complexity and large instrumental augmentation, Beethoven's storm has a greater impact because of its simplicity—with far fewer orchestral forces.

Having just noted Berlioz's reaction to Beethoven's storm, it is fitting that the second storm to be considered here was composed by Berlioz. *The Royal Hunt and Storm* (from *Les troyens* [1858]) opens with a forest scene that is shaded by dark clouds overhead warning of an impending storm. Horns delightfully tell us that a hunt is in progress. The images are vivid, as people give chase to the quarry, some of the chasers likely on horseback, others on foot, beating the bushes so to speak. Then, the rain starts; it comes in torrents. In the midst of the storm, various forest creatures dance wildly. The music reaches a huge climax as, for at least two listeners, a bolt of lightening seems to hit and uproot a tree, and we can imagine the forest creatures scurrying off and disappearing into the woods (Charles and Nancy Boody, personal communication, February 19, 1963).

The third storm never comes! The poetic third movement of Berlioz's *Symphonie fantastique* (see also pp. 99–101), entitled "In the Country," takes place one summer's evening. The young man central to the work wonders about his beloved as shepherds are heard calling each other. In time, following a repeat of a shepherd's rustic tune, the sun sets, and far away there is rumbling thunder, followed by silence. The storm lurks in the distance and then simply fades away. There is no grand explosion, just the hints from afar. The thunder is represented realistically by the use of four timpanists. They play quietly (single notes and rolls) interspersed with a clarinet's plaintive tune. The four notes (A-flat, B-flat, C, and F) on the timpani combine in an innovative manner to give interesting shading to the thunder that is so far away. As the thunder dies away, a horn and strings complete the piece with three bars of very quiet playing, followed by a second or two of silence, before the orchestra starts the next movement.

Beethoven and Berlioz drew upon a rich history of storms in music, as developed in the orchestral and operatic literatures (by, among others, Jean-Baptiste Lully, Antonio Vivaldi, George Frideric Handel, and Franz

Joseph Haydn), but they challenged that heritage. Previously, storms were represented by loud, fast "scrubbing" on the strings. Beethoven's storm set new standards in the establishment of what could be accomplished with sound using the standard orchestra. Berlioz also used the standard orchestra in *The Royal Hunt and Storm,* though his orchestra included the heavy brass, but in *Symphonie fantastique,* he did something different, creating a totally new sound with chords provided by multiple timpani. His image of thunder's complexity, set off in the distance, was also something quite new.

HUMANS AND NATURE IN REGIONS WITH EXTREME CONDITIONS

It is hard to live with only limited technology in harsh environments, as in Canada's North. The Inuit (known formerly by outsiders as the Eskimo) had a truly tough time living by "traditional" means (i.e., before guns and, later, snowmobiles were introduced and supplies were flown in from the urban-industrialized South to stores in their small settlements). The extended winter of dark skies for months on end and the varying weather changes influenced decisions as to when food could be hunted and gathered. Storms and shifting ice could prevent hunters from moving; hunger often resulted. Life was precarious. But the people knew their harsh environment intimately; they could "read" their landscape with superb skill. Inuit knowledge of and skill in reading their environment is expressed in the dozens of words in their vocabulary to capture varying snow conditions. Southern Canadians, in contrast, have only a handful of words to describe different snow conditions. Composer R. Murray Schafer compiled Inuit words for snow and wrote *Snow Forms* (1983), a fascinating musical rendition of the many words.

Travelers in harsh environments, such as those of Canada's North, especially in winter, must be cautious, for they may die if they misread the landscape and miss finding a settlement or essential resources. Inuit in Canada's North know their environment well and, so, are able to live in and travel across their outwardly barren land, as did their ancestors for thousands of years. In contrast, early European explorers of the Arctic and, later, the Antarctic regions had no experts to call on. They had to learn as they went about how inhospitable those environments can be. They sought to conquer nature, as evidenced by many failed expeditions to the North Pole and the first assaults on the South Pole.

Early in the twentieth century, a Norwegian team under Captain Roald Amundsen and a British team under Captain Robert Falcon Scott (1868–1912) both had the goal of reaching the South Pole first, with the

British team arriving to find the Norwegian flag planted in the snow. What a terrible blow it must have been for the five men of the British party to see the alien flag flapping in the wind where they thought they would be planting their own flag. The traverse Scott and his colleagues made to the South Pole was arduous. Dispirited, they set out on the return leg, to get once more to the edge of the ice shelf where their expedition's headquarters lay, at Hutt Point, McMurdo Sound, but they did not make it. The awesome trial of character that Scott and his colleagues underwent as they sought to overcome the challenges of the Antarctic have been captured in words, film, and music.

The 1948 movie *Scott of the Antarctic* dramatically tells the story of heroism defeated in the 1910–1913 British Antarctic Expedition, led by Scott. The movie is greatly enhanced by Ralph Vaughan Williams's incidental music. Deeply moved by the epic story, Vaughan Williams subsequently sought more fully to capture in music the powerful sense he had of humans pitting their spirit against uncompromising nature. Even as he wrote the music for the film, the idea of a separate symphonic work developed, and, in 1953, his Symphony no. 7, *Sinfonia Antartica*, premiered. This exploration of the Antarctic environment is fascinating, for the resulting sound pictures serve as an excellent example of how music can reveal a daring feat in a landscape of extremes.

The symphony is not as clearly programmatic as the movie score, but it nevertheless contains many descriptive passages reminiscent of that earlier score, of actual landscapes in the Antarctic, of an array of penguin, seal, whale, and bird life, and of human courage. The second and third movements are geographically descriptive, but sections of the outer two movements provoke thoughts of humans trying to overcome the desolate and rugged southern environment. While I listen to the music, I enjoy looking at the photos taken by the expedition's official photographer, Herbert Ponting (1921 and reprinted numerous times). Vaughan Williams saw his photographs. Ponting was not in the party Scott selected to go with him on the push to the South Pole, but his photos tell many things about both the Antarctic and the expedition. The juxtaposition of his superb photos with the music can move me to laughter or to sober reflection. Of note, the expedition included two geographers.

The expedition to the South Pole was made in the name of the King and the British Empire; hence, it took on values that gave it special meaning in the hearts and minds of those around the globe who owed allegiance to the King and valued the empire. Any spiritual importance to the trip for the participants was set aside for the perceived values inherent in privilege, honor, and allegiance. However, there also was the desire to vanquish nature by making the journey successfully to the South Pole and returning safely. I think Vaughan Williams felt that there was much more to the trip than just good, solid, British values; he saw the trip in larger terms, as humankind against nature, nature being all that the Antarctic could "throw" (almost literally at times) at them. Two ideas are thus explored: heroic adventure while seeking to overcome harsh nature, and the juxtaposition of nature's exquisite beauty and harsh ruggedness.

Vaughan Williams prefaced each of the five movements with short statements from literature. That for the first movement, "Prelude," is from Shelley's "Prometheus Unbound":

> To suffer woes which Hope thinks infinite;
> To forgive wrongs darker than death or night;
> To defy Power which seems omnipotent . . .
> Neither to change, nor falter, nor repent,
> This . . . is to be
> Good, great and joyous, beautiful and free;
> This is alone Life, Joy, Empire, and Victory.

The work opens with a sense of empty, white wilderness conveyed by a wordless female choir and a wind machine. Diverse ice forms are then brilliantly captured by the celesta, piano, glockenspiel, and vibraphone. The threat of wind, storm, crushingly cold temperatures, and ice movement are represented by themes played by the full orchestra that grate on the ear and gnash at the mind, with a bell signifying danger.

The scherzo second movement takes its theme from Psalm 104:

> There go the ships:
> there is that leviathan,
> whom thou has made to take his pastime therein.

As one listens to the opening horn motif, it is easy to imagine being on board the small ship (the *Terra Nova*, which had both a coal-burning engine and sails and sailed from Port Chalmers in New Zealand to McMurdo Sound) in dark, angry, churning waters. The voyage was rough for the passengers, crew, ponies, and dogs. (Scott was disappointed in his expectation that ponies would pull the sleds to the pole and back. Though trials with horses in Scandinavia were positive, the Antarctic environment was too tough for them. And though Amundsen and his colleagues used dogs—for their pulling power initially and as food later—Scott strangely chose not to do so. Hence Scott and his four colleagues—Edward Wilson, Henry Bowers, Lawrence E. G. ["Titus"] Oates, and E. R. G. R. ["Teddy"] Evans—set out on foot, dragging a large sled.) "Leviathan" appears out of the sea (played by woodwinds; the same theme in the movie referred to whales) as skua gulls and other birds screech. Weddell seals flop on the ice, where killer whales strain to reach them. There follows a hilarious musical scene in which we hear the picturesque, ungainly, waddling gait of the elegant emperor penguins—"courtly, with polished manners"—and the smaller adélie penguins—"busy, lovely little people" who are "the comedians of the South" (quotations from Ponting 1921, pp. 227, 229). The movement ends quietly.

The third movement, "Landscape," reveals an inhospitable landscape, but one with heavenly qualities revealed by exquisite sonorities that seem to hang in space as they evoke the images of a clear, windless day. In that scene are images of a wondrous grotto in an iceberg (photographed by Ponting 1921, p. 67, photo XXX), the ice home of echoes (photographed by Ponting 1921, p. 67, photo LXIX), the magic of the ever-changing clouds, and the beauty of the mountains. The musical moment is enchanting. However, that moment of reflection and respite is shattered by a huge blast from a solo organ, which abruptly and dramatically announces the existence of an impassable icefall (see photograph by Ponting 1921, p. 67, photo XXVI). The significance of the latter is explained in the third movement's opening quotation, from Samuel Taylor Coleridge's 1802 "Hymn before Sunrise, in the Vale of Chamouni":

> The ice falls! To that from the mountain's brow
> Adown enormous ravines slope amain—
> Torrents, methinks, that heard a mighty voice,
> And stopped at once amid their maddest plunge!
> Motionless torrents! Silent cataracts!

The fourth movement, "Intermezzo," provides some relief from the seriousness of the landscape portrayals, as human love is shown to contrast with heartless nature. Two lines from John Donne's "The Sun Rising" remind us of how important human love can be for those who struggle against nature: "Love, all alike, no season knows nor clime,/Nor hours, days, months, which are the rags of time." But then the scene is bleak; a blizzard is raging. The men are barely able to travel, and their food supply is almost gone. The tone turns darker still as the bell that was heard in the prelude is heard again, this time in remembrance of Captain Oats, who (because of weakness and pain from dreadful frost bite, hunger, and exhaustion) knew that he could not go on. His colleagues had been helping him move. Rather than continue to hinder their chances of getting to the desperately needed food supply, he walked out into a blizzard, never to be seen again.

The final movement, "Epilogue," draws together themes from various parts of the work as it works toward a resolution. This movement takes as its theme the final few words recorded in Captain Scott's diary: "I do not regret the journey; we took risks, we know we took them, things have come out against us, therefore we have no cause for complaint." The movement opens with a full-orchestral reference to the prelude, before depicting nature gradually overwhelming humankind. (Prevented from leaving their tent for four days by a terrible gale prior to their last day of life, they thus never made it to the cairn called One Ton Camp, just eleven miles away, where there was food and fuel. Exhaustion and their inability to walk any further due to conditions, overcame the men as they lay in their tent, and the tent was soon covered by snow. A rescue team set out from the coast when the weather cleared but arrived too late.) A bell soberly tolls as a solo soprano voice wails a wordless lament that is interwoven with windlike sounds by the women's chorus and the wind machine, surely signaling the loss of hope and then the loss of life. The voices and wind machine gradually fade to silence: nature has restored the Antarctic to its desolate emptiness. The silence seems to suggest that nature is all dominant; hence, we must ask, is there meaning to what we do, or are we doomed eventually to be overtaken by nature, despite our valiant efforts?

THE SAME ENVIRONMENT FROM ANOTHER PERSPECTIVE

Vaughan Williams wrote a partially descriptive work that is easy to follow as it unfolds. In many ways, the work speaks to a people's desire for heroism. Scott and his colleagues were long held by the British (and most certainly British youngsters) to be heroes. A quite different work approaches the Antarctic from a very different stance. It is nonprogrammatic; yet, at least partially, it is descriptive. Unlike Vaughan Williams, who never saw the Antarctic in person, the composer of this more recent work, Peter Maxwell Davies (1934–), benefited from having been able to visit that region, in the summer of 1998–1999. The outcome was the *Antarctic Symphony*, Symphony no. 8 (2000). The composer states that the work cannot be described literally in terms of a program associated with his Antarctic experience, for, rather, it is an abstract work that uses "transmuted sound images distantly based on those experiences." He draws on his own modified concept of time, which reflects others' research into tiny "unicellular creatures discovered in the rock below the polar ice, which lived for centuries" and the "relatively large sea-creatures whose slowness in metabolic rate and physical movement, and their long life span, are related to the restricted food supply and very cold water." In the music, layers of time are suggested, including even "its near suspension and the [to us] unusual rates of directly experiencing the breathing of time" (Maxwell Davies 2001, pp. 3–4).

Maxwell Davies opens the symphony with the sound of breaking ice as the ship carrying the composer enters the hostile environment. (For the ice-breaking effect, a percussionist shakes a metal biscuit tin filled with broken glass!) Later, a section makes references to some of his earlier work in order to suggest that they are consigned "to an ice-bound junk heap," this being a reaction to his dismay at seeing the debris left by years of activity by research teams. This strong reaction may be a reflection of his Christian beliefs and of his life in the timeless landscape of the Orkney Islands, where respect for community, the environment, and the past is vital and precious. Another section of *Antarctic Symphony* suggests the partial midsummer meltdown of icebergs, and the work ends with reference to "the harmonic essence, as if the ice has melted, revealing the rock beneath" (Maxwell Davies 2001, p. 4). A work clearly different from that by Vaughan Williams, it reveals in its unique way an understanding of the Antarctic based on site observations, rather than pure imagination and a story modified for a movie and thereafter remade as an orchestral work. As a geographer used to doing fieldwork, I am struck and impressed by Maxwell Davies's desire to complete a period of in-the-field observations before writing his score.

COMPOSING AND THE SEARCH FOR MEANING

Each of the works discussed in this chapter in some way deals with difficult environments, such as those based on folk beliefs (*Night on Bald Mountain* and the several works on the Faust theme), actual and potentially problematic storms (for instance, in the *Pastoral Symphony*, when the people at play must run for shelter when Beethoven's storm arrives), volcanic situations, expressions of heroism and the desire to conquer nature (*Sinfonia Antartica* by Vaughan Williams), the harming of the Antarctic environment (which had claimed the lives of Scott and his men) due to a more advanced level of technology than any available to the Scott expedition (*Antarctic Symphony* by Maxwell Davies), and individual's searches for meaning (Hovhaness, for instance). To extend this discussion, with an emphasis on composing and the search for meaning in relation to the sacred as evidenced in nature, I now focus on a composer who received a passing mention earlier in this chapter.

Gustav Mahler's work, possibly more clearly than that of any other composer, reveals what a composer—or, indeed, a serious thinker—must go through to gain an understanding about relationships to the sacred in nature. Mahler revealed his agonies, searches, findings, and remaining questions in his work. His Symphony no. 1 (1888) (known as the *Titan*) opens with a wonderful evocation of nature, specifically spring, the first of many references in his ten symphonies to his interpretation of pastoral scenery. Indeed, references to nature abound in all of his symphonies, but married to his landscape interpretations were his personal explorations of the purpose of life. For example, his Symphony no. 2 (1894), known as the *Resurrection Symphony* (though Mahler did not name it thus), explores this issue; more pointedly, he asks, "To what purpose have I put my life?" The work is nothing less than gigantic. It calls for a much enlarged symphony orchestra, but size does not in itself necessarily mean loud, expansive sound in his work—in contrast to Strauss's. Under skilled conductors, Mahler's huge orchestral forces can be powerful in expression yet marvelous in constraint. The lengthy first movement of *The Resurrection* began as a tone poem, *Totenfeier* (*Funeral Rites*) (1888), but was developed to express more clearly the concept of the life force's ability to rise from the ashes of fate through faith in God. The second and third movements are retrospectives, the second "being innocent and nostalgic," whereas the third includes a "certain element of the grotesque" (Steinberg 1995, p. 287). In the fourth and especially the (very long) fifth movements, the work's hero hears the call of God, but he then must endure the Day of Judgment before being granted redemption, resurrection, and immortality. It may seem strange that Mahler began the work with a funeral march, but, of course, he included funeral marches in six of his ten symphonies. Use of these marches surely reflected his own uncertainty about life.

In *The Resurrection*, as in all of his works, Mahler uses marked disjunctions of tempo and sound. This is dramatically evident in his almost miraculous last trumpet call—*die grosse Appell* (the great summons)—in which trumpets and horns call out from afar (the players are generally located offstage), while silence descends as a solitary bird (oboe) flies around the landscape of destruction. A hymn of resurrection emerges from the silence. The beautifully pure sounds from the voices are joined by several instruments, and then, as the word *rief* (called) is sung, a soprano soloist floats free of all else, with immense beauty and meaning. Indeed, this is one of the great moments of consequence in music. The work, powerful from stem to stern, ends in a celebratory, clamorous fanfare and ringing of bells.

In a sense, the *Resurrection Symphony* is a grand hymn to humanity's salvation through spirituality, yet it does not stand alone, for Mahler's remarkable *Symphony of a Thousand*, Symphony no. 8 (1907), expresses the power of the transitory, while drawing upon the intransitory, the "eternal bliss," that "lies behind all experience, that is indescribable [but] which draws us by its mystic force [and] which every created thing, perhaps even the stones, feels with absolute certainty at the very center of its being" (Mahler to his wife, June 1909, quoted in Steinberg 1995, pp. 337–38). Mahler was very much taken with the example and significance of Beethoven's Symphony no. 9, and he copied Beethoven's use of a chorus.

Within his symphonies, and notably in Symphony no. 2, Mahler generally stated themes early and then elaborated upon them to achieve an ever-changing tapestry of sound reflecting constant reinterpretation. His conclusion seems to have been that we can go back to earlier thoughts and experiences, see them anew, and grow from them. His spiritual search as conveyed in his many musical canvasses is both persistent and demanding, with remarkable effect on any listener.

The point of identifying Mahler in this manner is to suggest that all composers, whether they reveal their inner thoughts in music or not, have to struggle. Especially noteworthy for our purpose is the point that, as Mahler composed, he drew upon the landscapes around him for comfort and inspiration. And, of course, we benefit due to the expression in his music of nature—and of his personal search for meaning.

EARTH MASS

The earth, with some landscapes of extremes, is also the planet that gives humankind much to rejoice over; therefore, let us conclude this chapter by considering, first, *Missa Gaia*, and, second, some of Einojuhani Rautavaara's music. A black mass was performed in Mussorgsky's *A Night on Bald Mountain* to celebrate evil. Rather than letting evil have the last say, let us recall that Hovhaness's broad ecological and spiritual approach to nature led him to appreciate and celebrate its wonders. The next work takes a similar stance to that of Hovhaness as earth is celebrated in all of its goodness. Accordingly, if the devil can be given his expression in music, then we must consider other celebrations, too, and one such is provided by Paul Winter in *Missa Gaia/Earth Mass* (1982), a "mass in celebration of Mother Earth" for chamber orchestra and choir. Partially based on the form of a Christian mass, it angles off in many ways to incorporate both words and music that are anything but close to what may be regarded as standard. The mass opens with a song based on some writings of St. Francis of Assisi, patron saint of ecology, who called many creatures, including sheep, waterfowl, birds, crickets, and wolves to give praise and prayer to the Creator. Thereafter, in the "Kyrie," focus is given to a female tundra wolf. "Beatitudes" is followed by a musical rendition entitled "Cathedral Forest." Next, an improvisation for organ and soprano saxophone evokes a dream fantasy of seeing Gaia, the Earth, from space. Three pieces follow, to form a whole: an instrumental anthem, which includes the throbbing "Heartbeat of Mother Earth," represented by a large ceremonial drum and the song of a Grand Canyon wren; "Adoro te devote," a thirteenth-century Gregorian plainchant; and, as the latter song has come to the present, "For the Beauty of the Earth." A John Whittier poem, "Sound over All Waters" leads to the "Sanctus" and "Benedictus," which builds on the belief that the whale is jubilant. The whale that inspired this thought was recorded off Brazil's northeastern coast, and the music incorporates the joyous rhythms of the *Baiao* from that region. "Ube caritas" (Where There Is Love), using a blend of Gregorian plainsong and African chant, leads to a cello improvisation, "Stain Glass Morning," that was inspired by New Zealand's Bellbird and the Amazonian *Uirapuru* (musician wren). Other pieces follow, including one that uses a conch shell to open a celebration of the Brazilian goddess of the waters, before the departure introduced by a trilogy of voices from the wild: a loon, a wolf, and a whale. The integration of music from various sources and the many references to nature, patron saints, and goddesses of nature reveal an interesting progression away from the mass as celebrated in Christian churches, despite being based on the form of the mass. This work of worship creates a fascinating landscape of blended sounds from nature and many cultures.

NIGHT SKIES

This chapter has considered extreme environments and has demonstrated that meaning (for good or evil) can have landscape associations. Following the joy and peace of Paul Winter's celebration, let us turn to the skies—not skies conquered by telescopic and capsule penetration but those captured in music. I write about this because of a memorable evening. I first heard Symphony no. 7, *Angel of Light* (1994), by Rautavaara (1928–) while at my cottage in Vermont. It was a fall evening, the sky was clear, and stars were abundant. As I heard the quiet opening section of the work, I could see myriad stars twinkling above me. Then, as if on cue, I saw the wonders of the aurora borealis start to dance across the sky to the north. As the music continued, I saw an ever-changing, wondrously stunning display of light, rarely staying still, moving into new shapes and colors as if they were rapidly moving, thin veils of cloud. The feeling evoked within me was one of wonderment and delight. The juxtaposition of the "light show" and the sparkling music was truly magical.

7

Landscapes of Death, Survival, and Remembrance

VIOLENT DEATH AND CONSEQUENCES

The human condition being what it is, conflict occurs, all too often, with fighting, fear, horror, loss of life, and a generation of despair. Armies of old killed, plundered, and raped; modern armies have been known to do the same. Civilians always lose. Out of war and its related turmoil, survivors may have such a strong desire for revenge that old enmities are kept alive, even when what is needed is a time for mourning, hope, and reconciliation. How have composers dealt with these matters? Some composers share their anguish by revealing their deeply disturbing insights, while others take a detached and heroic stance.

The theme of this chapter is derived from the way I think as a cultural geographer. As in other chapters, my exploration of the identified theme is based on issues that emerged from reading dozens of compositions and associated literature. The issues I ultimately identified are untimely death, depositing the dead, battles and victory, anger at wanton destruction of and threats to life, the need for remembrance, a call for reconciliation, funeral marches and requia, music in heaven and purgatory, the importance of music for cathartic purposes, and loss of freedom. We begin with a landscape of remembrance, following a violent death.

SILENCE, THE MILITARY, AND A FUNERAL MARCH IN A LANDSCAPE OF REMEMBRANCE

Particular moments in history can be ingrained in the memories of millions of people. If music is involved, it may trigger recollections. By way of an example, Washington, D.C., was a quiet landscape of mourning two days after President John F. Kennedy was assassinated on November 22, 1963. The capital city's public landscape was remade in the image of a national hero, but not by things constructed for the event—not even barriers to keep people back as they stood along Pennsylvania Avenue and other roads—but by the simple gathering of hundreds of thousands of people who watched the thirty-two men in impressive dress uniforms (there were no women) escorting the president's coffin to the Capitol. Other than any color on the uniforms, the scene was essentially grey, not least due to the leafless trees. President Kennedy's black caisson was drawn by six white horses, four of them mounted by military officers. The crowds that day and the two to follow were silent. The procession, "with a slow and measured tread" (*Four Days* 1963, p. 74), wound its way from the White House to the Capitol, accompanied by the corps of military men and twelve drummers, who, by tradition, played muffled snare drums. The steady drumbeat became like a tattoo in the mind. After lying in state under the Capitol's rotunda for a day, the coffin was taken in another small and simple procession from the Capitol to the Roman Catholic cathedral for a pontifical requiem mass. Then, joined by the family and hundreds of foreign dignitaries, the caisson, again pulled by the white horses, moved to the burial site in "the haven of heroes" (*Four Days* 1963, p. 119), the military cemetery in Arlington, Virginia. On that final leg of the trip, many a tear was shed at the sight of a riderless black horse as it was led behind the casket. Backbones quivered when the military band played Frédéric Chopin's *Funeral March* as the cortege came close to the White House. As the last trumpets faded, that haunting music was replaced by the continuing silence of the crowds, the clanking of the cortege, and the sounds of horses hooves, muffled drums, and soldiers marching. Chopin's tune was repeated again and again on radio and television stations for four days, from the time of the assassination until the burial. Thus, Chopin's *Funeral March* became indelibly ingrained in the minds of countless numbers of people. It still triggers memories of that day. Other music included "Hail to the Chief" (the song of recognition for the president of the United States) and "Eternal Father, Strong to Save," the U.S. Navy's official hymn.

Many composers dedicated their compositions to Kennedy's memory. Canadian André Prévost's *Fantasmes* was premiered by the Montreal Symphony Orchestra on the very day Kennedy was murdered. Subsequently, Prévost inscribed the score: "In memory of John F. Kennedy, president of the United States, victim of the world which I have described here in my music." Igor Stravinsky asked the British poet W. H. Auden to write the words that Stravinsky then put to music as *Elegy for J. F. K.* (1964). Their message? It is up to us how well Kennedy will be remembered.

DANCE OF DEATH, THE GUILLOTINE, AND THE CATACOMBS

For piano and orchestra, *Totentanz* (1862) by Franz Liszt, is powerful for its macabre nature. Various attitudes are conveyed during this "dance of death": mockery, triumph, contempt, reverence, compassion, cynicism, and, ultimately, resignation to the fact of death. A conductor once told me that he imagines the devil dancing whenever he hears this work.

French composer Francis Poulenc's (1899–1963) *Les dialogues des Carmélites* (1956) is not a dance of death, but an opera, based on a true story, with one of the most startling and moving scenes involving death. Its impact is dramatic. Set during the French Revolution, we learn that priests are to be denied the right to give Communion and then that the king is dead. Some Carmelite nuns take a vow of martyrdom, though one flees. The nuns are jailed, convicted of treason against the republic, and sentenced to die on the guillotine. In the Place de la Révolution, one by one, the nuns are killed. They sing as they await their turn. (But of course, this is opera!) The chorus gets smaller and smaller until there is one nun left. Just before she, too, is murdered, the nun who had run off earlier appears. They recognize each other, and the singing resumes for a few more bars, but suddenly stops when the nun on the scaffold dies. At that point, the "renegade" nun steps out of the crowd and moves toward the scaffold, singing in defiance. She sings the final four verses of "Veni creator," but then her singing suddenly stops as the blade drops. The power of the music in this final scene is awesome.

Mention of this French work brings us to the issue of where to put bodily remains once cemeteries are filled. There is no denying that in due time we all die, and our bodies have to be dispatched in one way or another. Landscapes of death in many cultures involve places for burial, cremation, or some other means for disposing of corpses. Because cemeteries leave clues about former life and societal matters, it is useful to think of them as *living* landscapes (Knight 1982a, 2005). In some urban places, bodies are buried for a set number of years and then must be dug up so others can use the space. The problem of what to do with the dug-up remains in Paris was solved by putting them even further beneath ground. The Catacombs beneath the Fourteenth Arrondissement in Paris are one of the wonders of the humanized—or should we say dehumanized—landscape. In that chaotic "empire of the dead" are to be found the undifferentiated remains of six million humans. Skulls, femurs, and tibias are piled high, as if silently awaiting Judgment Day; it remains to be seen what they will do without their full bodies when that day arrives. On occasion, these human remnants have been blessed with music, as on April 2, 1897, when forty-five musicians secretly descended into the catacombs and proceeded to play Chopin's *Funeral March*, Ludwig van Beethoven's *Funeral March* (the slow movement of his Symphony no. 3, *Eroica*), and Camille Saint-Saëns *Danse macabre* (Gup 2000, p. 144). Not merely an occasional site *for* music, the catacombs are also set *in* music. Modest Mussorgsky created in music a brief, but vivid, image of the catacombs (see p. 102). The section of *Pictures at an Exhibition* entitled "The Catacombs" conveys a sense of gloom, perhaps due to poor light from candles, but also surely due to the millions of skulls "staring" at nothing. What is the purpose of life, one asks, but gets no response. This is a dead landscape.

The idea of bringing skeletons "to life" obviously tickled the fancy of French composer Saint-Saëns, for in his symphonic poem *Danse macabre* (1874), there is a joyous rendition of skeletons (the xylophone) leaving their graves at midnight and, with bones rattling, dancing in the moonlight, while the orchestra plays a morbid marchlike tune. The image of the interplay between the orchestra's doleful manner and the mischievous dance is readily understood. The fascinating scene ends when a rooster crows the signal for morning's arrival.

WAR AND TRIUMPH

Many deaths result from war. Englishman Gustav Holst's *The Planets* (1918) opens with "Mars, the Bringer of War." It opens quietly, but threateningly, as all the strings, two harps, and the timpani (with wooden sticks) beat out a steady rhythm that grows in a gradual crescendo. Hints of threat are given by the brasses and strings, but then, at bar forty, there bursts forth a mammoth outpouring (*fff*) of the basic repeated pattern on strings, trombones, tuba, and timpani; then, with the timpani, trumpets and strings continuing the menacing pattern, the deep brass and horns pronounce a threatening presence. The interplay continues, and the music drives on (with all instruments reaching for an impossible *ffff*). Holst's message is clear: war is evil; be aware, and be prepared. Holst wrote this movement in 1914; hence, many in Britain saw it as a harbinger of what was to come: the cataclysm of World War I.

Music sometimes finds its way onto the battlefield. The Scots were renowned for using the bagpipes—sometimes a solo piper—to strike terror into the hearts of adversaries as the troops advanced on enemy lines. Since I love the pipes, and as a young drummer played in pipe bands, I cannot understand such a reaction, but this simply suggests that familiarity and perceptions play a role in musical appreciation. Of course, too, it cannot be refuted that battle conditions are a factor, with one side stirred on by the music, the other filled with increasing apprehension as the piper gets closer.

Musical battle landscapes have been created by a number of composers. One with little expression of consequences for those involved is Beethoven's *Wellington's Victory* (1813) at the battle of Vitoria, when British forces beat Napoleon's troops. By way of warning, even in his own estimation, this is not one of Beethoven's better works. To many listeners, it is thoroughly enjoyable, but it is corny to others and hated by a few. Beethoven wrote into his score when rifles (ratchets) and cannons (bass drums) are to be fired. The orchestra sits onstage in its normal seating arrangement, but at Beethoven's direction, two sets of trumpets and drummers (one group British, the other French) are lined up on each side of the front of the stage "firing" at each other. The effect for the audience is exciting. The British, on the right, start. A drum tattoo (Beethoven checked for accuracy) can be heard faintly in the distance, waking the troops. The volume increases as additional drummers join in. Out of the din, with all British drums rolling, trumpets sound a battle cry, followed by a rousing rendition by the orchestra of "Rule Britannia!" From the left, the French respond in kind (the drum tattoo being accurately French), only their rousing orchestral tune is "Malbrouk s'en va t'en guerre." A brilliant French trumpet call challenges the British to fight; the British accept, throwing back the call. The battle begins, with the full orchestra joining in. Guns and rifles "fire" across the stage—just missing the conductor—at the exact times Beethoven wrote in the score; blocks of troops enter the fray; and, conveyed by music of an insistent tone with a constant rhythm played at increasing tempos, the cavalries charge. One can almost smell the gunpowder and sweating horses and hear the screams of men and horses being killed, though the emphasis in this music is on triumph, not pain. The French defeat is signaled by the playing of the British national anthem, "God Save the King," and the work then comes to a triumphant conclusion.

Beethoven revered Napoleon for a long time; however, when Europe witnessed Napoleon letting power go to his head, as it were—for he was crowned King in May 1804, upon his own order—Beethoven was bitterly disappointed. He had already written a dedication to Napoleon onto the original score of his Symphony no. 3, but in a fury (judging from the mess he made as he erased the inscription), he scrubbed out Napoleon's name, dedicating it instead to Prince Lobkowitz (a patron of Beethoven), and in 1806, a year after its premier, named it *Sinfonia Eroica* (with his inscription on the first printing being, somewhat mysteriously, "a heroic symphony . . . composed to celebrate the memory of a great man"). The *Eroica*, as it has been called ever since, is arguably the greatest symphony ever written because of its innovativeness, driving energy, and majesty. It is narrative in a sense, thus anticipated the type of programmatic music soon to be developed by other composers, but the extramusical references are more elusive in this work than in *Wellington's Victory*. The symphony conveys images ranging from abject misery to pure exultation; such an array of emotional insight was totally new in music. The first movement is exciting but seemingly without references we would recognize as either landscape or "hero" based. The second movement, however, is a funeral march for the hero; it has great emotive power. After a stunning scherzo filled with near bursting energy, the finale provides (what to Beethoven's original audiences was) an innovative exposition and resolution. Among other things, the finale cites English and Hungarian music by way of acknowledging the cooperative effort of a number of European states that fought together to defeat Napoleon's growing imperial aims. It resolves many inner questions and tensions within the score by releasing all in a wondrous conclusion. This symphony is thus different from the nonsubtle *Wellington's Victory*, in which the story line cannot be missed. *Eroica* has subtlety, complexity, and power, which together move the spirit and challenge the intellect. Beethoven's Piano Concerto no. 5 (1809) is sometimes called the *Emperor Concerto* to imply Napoleon, but this tag is totally inappropriate. As D. F. Tovey has stated, it was named *Emperor* to Beethoven's "profound if posthumous disgust" (quoted in Steinberg 1998, p. 71).

Zoltán Kodály in "The Battle and Defeat of Napoleon," part of the *Háry János Suite*, turns from beauty in the dreamlike previous section to the grotesque. Using a combination of winds and percussion, however, the orchestration suggests that the battle scene should not be taken too seriously. The opening march gets fuller as more instruments are added, but, then, with a wave of the conductor's baton, the French emperor's army falls "like tin soldiers" (Breuer 1990, p. 108). The supercilious Napoleon is brilliantly portrayed by the bass trombone and tuba. The section ends with a short funeral march, noted for its saxophone solo.

After victory, surviving troops go home. In the last section of *The Pines of Rome* (chapter 4), Ottorino Respighi has Roman soldiers marching on the Appian Way, indicated by a steady timpani beat, as they approach Rome. The sense of triumph is pungent as the music rises to a huge climax. As does Beethoven in *Wellington's Victory*, Respighi creates a sense of movement that is underscored with tension, anticipation, and triumph. Both composers and their interpreters—the orchestra players with a conductor—successfully create for the audience mental images of particular places and events.

Before victory—or defeat—soldiers march, ride horses, or, in the last century and today, drive tanks or other armored vehicles. Military battalions worldwide have songs and marches written to celebrate campaigns, with possibly the British army being the best known for its numerous celebratory tunes, played by military bands. Imperialism with a song! Eric Coates wrote the music for the film score for *The Dam Busters* (1954)—about the true story of daring British fliers who bombed three dams in Germany during World War II. The score includes a popular tune by the same name that now stands apart as a concert piece. It reminds us of air campaigns, but it is rather jolly. It does not evoke images of war or of tragic loss among the aircrews or, worse, of life downstream from the dams. The same is true of William Walton's heroic sounding prelude and fugue *Spitfire* (1942) from the film score for *The First of the Few*, so named for the famous World War II British fighter plane flown by the "famous few" in the Battle of Britain. On the same topic, Walton also wrote *The Battle of Britain* (1969/1985). During the nineteenth century, numerous composers included references to particular battles in their music, including Pyotr Ilich Tchaikovsky, who composed the *Battle of Poltava* (part of his music for the play *Mazeppa*) and the *1812 Overture*.

Written in 1881, the *1812 Overture* is pure Tchaikovsky, and because of that, stereotypically Russian. It has very noisy, yet brilliant, brass playing, thunderous cannons, and loudly tolled bells (representing Moscow's church bells) to celebrate the Russian rout of the French. An Orthodox hymn and the Russian national anthem are juxtaposed with "La Marseillaise." The image of the relieved and ecstatically happy populace in Moscow is inescapable. I once was astonished during a U.S. holiday to hear a television announcer say that the Boston Pops Orchestra was to play "that wonderful *American* celebratory tune, the *1812 Overture*" (my emphasis). Russian listeners must have shuddered. Even so, the work is played around the world with relish, for it is a joyous piece. There sometimes can be surprises in music. A violist friend once directed his brother to sit near a fence during a summer outdoor performance that included the overture. What he did not tell his brother was that the cannons (whose firing Tchaikovsky scored) were directly on the other side of the fence! I am assured that they are now again talking to each other.

Franz Liszt (1811–1886) used the Hungarian cimbalom to evoke a patriotic response to his *Hungarian Attack March* (1875). The work opens threateningly in a manner closely reminiscent of Beethoven's *Wellington's Victory*—even to the use of high strings and a piercing piccolo—but then resorts to an almost serene tune that is repeated, along with intermittent reminders of threat. It is almost as if the soldiers are out for a Sunday afternoon ride, only to be reminded occasionally that they are on serious business. This short work ends on a triumphant note. Is war always thus? We will shortly see that it is otherwise. Liszt also wrote the somewhat longer *Battle of the Huns* (1853?), a work inspired by a mural of a fifth-century battle between the Roman emperor Theodoric and Attila and his Huns, with the chorale "Crux fidelis" representing the Christian victory. Nikolay Andreyevich Rimsky-Korsakov's "Massacre at Kerzhentz" from the third act of the opera *Legend of the Invisible City of Kitezh* (1904) portrays the deadly conflict between soldiers of the city and the invading Tartars. Interestingly, Rimsky-Korsakov's opening bassoon solo, repeated by the clarinet, and the staccato sounds he creates with the whole orchestra seem surely to have been a source of inspiration for Dmitry Shostakovich many years later. Dramatic as these various works are, we turn to World War II for some powerfully expressive landscapes of war.

THE HORROR OF WAR

Dmitry Shostakovich (1906–1975), one of Russia's most innovative composers, had to walk a fine line between self-expression and meeting the demands for conformity identified by the Soviet "establishment" (Stalin and his bullying henchmen who controlled life in the Soviet Union, including leading musicians who toed the party line and, in turn, bullied Shostakovich). He is controversial because he seemed at times to be a conformist, and, yet, he was a master of double meanings. Such is the case with his powerful Symphony no. 7, called by others the *Leningrad*. Shostakovich was living in Leningrad (today renamed St. Petersburg) during the early months of the Great Patriotic War, as World War II is referred to in Russia. Despite a recently signed nonaggression pact between Stalin and Hitler, Hitler sent his troops into the Soviet Union in 1941, the intent being to take Moscow quickly. As is known, the Germans under Hitler repeated the experience of the French under Napoleon a century and a half earlier: defeat, largely due to the weather, poor supply lines, and inadequately trained, clothed, and fed troops. The winter of 1941–1942 was grim in Leningrad, where the Germans imposed a siege that lasted for nine hundred days. About one million people, one-third of the city's population, died during months of shelling, fire, disease, cold, and starvation. Conditions were unbearable; yet, a valiant spirit of survival kept the city's inhabitants going. A simple monument in St. Petersburg stands in remembrance of those who lost their lives during the siege.

The dates noted by Shostakovich on the manuscript copy of each movement show that the work was completed during the first year of the siege, though for some of that time Shostakovich was elsewhere. After serving as a fireman in Leningrad, he and his family—and many other artists, notably from Moscow—were evacuated to Kuibyshev, the temporary capital of the Soviet Union due to the German advance. The symphony was completed and given its first performance in Kuibyshev in February 1942. It was played in Leningrad in August of that year by an orchestra of half-sick musicians (who were given extra rations just prior to the performance). The score was smuggled out of the country on microfilm to London (where it was performed to acclaim under the direction of Sir Henry Wood) and New York (where, a month later, Arturo Toscanini and his famous NBC Symphony Orchestra performed it, broadcast across the United States). The work captivated people abroad, being performed numerous times for several years in Allied countries, and it was a patriotic rallying cry for Russian audiences. Though at times, especially in the first movement, the work is powerfully descriptive, Shostakovich had mixed feelings about music that had extra musical meaning. Possibly because of that, he dropped the titles he had planned to give to each of the movements. Even so, the work retains its place as an awesome piece written in the face of destruction caused by a seemingly unstoppable enemy.

The title intended for the first movement was "War." At the outset of the movement, deep strings, woodwinds, and timpani, later picked up by the brass, convey a sense of threat, but this soon passes as a flute solo, followed by other woodwinds and strings, suggest that the inhabitants of the city and surrounding countryside are not yet fully aware of the danger that is advancing on them. The basses again identify the threat lurking in the background, but this, too, is ignored as the woodwinds and strings continue to sing, though they grow increasingly melancholy and then become silent. Have they, in Leningrad, heard something? What is it? In the distance, we hear a very quiet snare drum playing a march rhythm, to be joined ever so quietly by the violins and violas tapping out a staccato beat and a flute playing a tune that is almost pleasing to the ear but within which a villainous threat lurks—double basses and bassoon. The tune is passed around the orchestra, and bit by bit, over many minutes, it grows (like Maurice Ravel's *Boléro*), until it is clear that this "force" is relentless in its advance. This army will stop for nothing. The drum continues, louder and louder (the composer calls for two or three drummers by the end of the huge crescendo), as the brass and timpani join in the identification of the army's advance, tearing up the land as they travel in their tanks and trucks. Behind the basic beat, the high string's scream, surely representing the people in the countryside who burned their houses and crops and then tried to flee before the arrival of the Germans, though some of the screams could represent the people trapped in villages and towns by the advancing army. The screams could also represent the inhabitants of Leningrad upon realizing their fate. The army's march is unremitting. Leningrad is the goal. These two themes continue, the advance and the screaming, until, finally, the advance wins out in a dramatic close; however, Shostakovich surprises us, for this is not the end. Instead, the music continues quietly, periodically punctuated by the march in the background. This concluding section suggests that the people of Leningrad are now settled down for the duration of the siege. They are burying their dead, building their long fortification, manning lookouts, and, very basically, surviving. The images of a landscape trampled over by the advancing army and the powerful sense of place among Leningradians as they defend the city are palpable. Ironically, the Bolshevik Central Committee attacked Shostakovich in 1948 for having portrayed the enemy so well while not conveying even better the "heroic" spirit of the Russian people. Ideologues!

The short second movement was to be called "Memories." It is dance-like, but plaintive in character, played mainly by the strings and woodwinds (especially oboe with some rich, dark moments of bassoon). Quiet reminiscences, or memories, suddenly give way to an agitated recollection, perhaps to reflect the dangerous situation the country is in, but this is pushed aside as calm returns, with bassoon and harp and the return of the strings, by way of accepting that a grounding in memory will be important in the difficult days ahead. Shostakovich called the third movement, an adagio, "Our Country's Wide Spaces." He later revealed that he sought to convey a feeling of Leningrad at sunset, the streets and the banks of the river Neva suspended in stillness. It is perhaps an elegy to the city. The humanized landscape is almost serene. It does not remain quiet, however, for a sense of movement soon develops that is quite extraordinary, as Shostakovich goes without a pause into the final movement. Originally titled "Victory," this movement is dramatic. Clearly, Shostakovich was not saying that the Russians had won the war, for the siege was still ongoing, but perhaps he was suggesting the Russian people had already spiritually beaten their adversaries due to their resistance and perseverance. The underlying theme is dark: the strings convey this sense of darkness, underscored by a long timpani roll. Momentum builds as richness of sound and variety of melody convey a sense of victory, but this is reined in as we are reminded, in a slow, dark way, of the tragedy of countless deaths. Once this statement is made, the music pushes to an affirmative conclusion: though it has not been easy, it seems to say, the Russian people have indeed won.

Symphony no. 7 conveys an array of human emotions. The tension created between the advancing army and the screams of pain will surely remain with any listener. At the start of this discussion, I noted that Shostakovich was a master of double meaning. Perhaps the symphony was not telling the story of the invasion by the Germans but, instead, that of the crushing of the Russian people by the tyrannical power of Stalin and his henchmen—and of the people's resilience? Could it be that he intentionally had a double meaning, for each possibility is plausible? Perhaps not, since he did dedicate the symphony to the city of Leningrad, and, as noted, he had names for the various movements, facts later underscored by his son, so maybe that is all we need read into the work.

The possibility of double meaning also exists with his Symphony no. 8 (1943). Does it, again, refer to external war or war on the Russian people from within? Remember, millions of people were relocated, most to camps in the east, and millions more were starved or murdered by the Communist regime. Shostakovich was sensitively aware of the evils of the system he lived in, but he could comment only through his music. His symphony certainly conveys a sense of the horror of war in a truly bleak manner. Laurie Shulman has described the symphony as one of Shostakovich's most personal political statements. "The Eighth Symphony is the voice of an entire people engaged in mortal conflict. It addresses both the psychological trauma and the savagery of military combat." She continues, "Few symphonies by any composer have the unrelieved grimness and dark view of the human condition of the Eighth. This is a work that drains the listener's emotion" (Shulman 1997, pp. 3–4).

Shostakovich's Symphony no. 11, *The Year 1905* (1957), is yet another description, and indictment, of the horror of violence. Purportedly to celebrate the fortieth anniversary of the October 1917 revolution, it tells the story of unarmed citizens outside the Winter Palace on January 9, 1905, who were massacred by the Tsar's police. There are no breaks between movements. The opening movement moves slowly, starting with strings and harp, signifying the start of a day that is to be remembered with shame. A trumpet sounds. The day develops, with strings underscored by timpani, and then a horn recalls the trumpet's call. Various instruments "talk," generally with quiet timpani strokes hinting to the listener of the dreaded moment ahead. In due course, the crowd gathers quietly; the orchestra conveys a strong sense of foreboding. This leads into the second movement in which there is a long, slow passage but which gradually builds in tempo and, eventually, intensity (with rushing strings, brass choruses, timpani, followed by snare drum and trumpets, then double basses, horns, trombones, timpani, and strings, and then others, including a piercing scream on the clarinets, until the whole orchestra takes over), as the mounted police enter the crowded square. The awful moment arrives when the police open fire on the crowd and cut many people with flashing sabers (dramatically portrayed by timpani, bass drum, snare drum, gong, cymbals, and xylophone, undergirded by the strings, woodwinds, and strong brass); many die. Silence suddenly descends, and in the third movement, the music mournfully reflects on those who have lost their lives. The final movement revisits themes used earlier in the work, moving from haunting moments to dramatic tension and back again, and ends with a loud outpouring of anguish. Again, it is possible to appreciate that Shostakovich had a double meaning. While the "story" has a tie to an event in Russian history, the work was written shortly after Russian troops violently put down the 1956 uprising in Hungary, so it is possible that instead of the 1905 massacre, Shostakovich was thinking of Soviet tanks firing on unarmed Hungarian civilians. Lev Lebedinsky, a leading Russian musicologist and friend of Shostakovich, considers this interpretation a possibility, but Shostakovich is not alive to confirm or deny its accuracy, so we are left with the question. Whether or not there are double meanings, Shostakovich sought to tell the truth as he saw it. He was not a conformist, but a reactionary, to expected styles, as he demonstrated from the outset when, as a student, he wrote his Symphony no. 1, which said, essentially, I will be different. With reference to Shostakovich and two other composers, Michael Hall's words are important: "Truths can be expressed in music which, if written in words, would have cost the authors their lives" (Hall 1996, p. 123).

OTHER LOCATIONS, OTHER COMPOSERS

Fighting occurred elsewhere during World War II, of course, and other composers also wrote music. Nothing quite hits the mark like Shostakovich's music in terms of the revelation of the terror of war, but some other works are nevertheless significant, including those by Americans Harl McDonald (1899–1955) and Richard Rodgers (1902–1979). McDonald wrote a tone poem entitled *Bataan* (1942) to remember the stand in 1942 on the Bataan Peninsula of American and Filipino armed forces against the strong, invading Japanese army, which ended in victory for the latter. Rodgers (with arranger Robert Russell Bennett), in contrast, undertook to write the score for a series of twenty-six half-hour television programs entitled *Victory at Sea* (1952–1953), containing footage taken by U.S. naval personnel in both the Pacific and Atlantic theaters of war. The accompanying music conveys the essence of the many settings, events, and, to a degree, observers' feelings. Subsequently, a concert suite was made from the film score. The music varyingly conveys feelings of coldness, bleakness, loneliness, danger, playfulness, and, ultimately, triumph; however, Rodgers's work is not introverted. The opening movement, "The Song of the High Seas," is the recurring theme throughout. One thinks of the huge expanses of ocean and tiny ships, of calms and swells, and of the sea's ever-changing moods. The nine-movement work, *Victory at Sea*, includes music that represents, among others things, a German submarine attack on an Atlantic convoy, the attack on Pearl Harbor, a U.S. carrier task force and assaults on Micronesian islands, the power of gun fire, sailors' horseplay, a burial at sea, a convoy in a storm, and a concluding hymn of victory. The richness of the score is quite unlike that of Shostakovich's, perhaps because Rodgers wrote music to accompany a film "canvass" created as both a record of events and a work of nationalistic propaganda in which the audience could see all. His compositional voice and vocabulary reveal a tapestry that has far fewer raw edges than does Shostakovich's music. Rodgers, who lived in a "free" society, was openly patriotic and imbued his work with no hidden meanings. This noted, it is not clear how much, if any, pressure he was under to compose music that fitted the ideal the U.S. Navy sought to uphold in the huge film for which the music was originally written.

As World War II proceeded, the jangle and screeching of tank tracks, the roar of jeep and truck engines, the drone of war planes, whether fighters or bombers, the buzzing of V-1 and V-2 rockets fired toward London, and the dreadful concussive "thud-boom" of bombs exploding were all part of the soundscape of war (on the bombing of cities in Germany, Britain, and Japan, see Hewitt 1997a). Another part of the World War II soundscape was provided by the radio: "da da da daaa" was immediately recognizable everywhere as the call of the BBC, "This is London calling." That little motif of four notes was the sound of hope for many people trapped in war. For some, the BBC conveyed "just" news, but for others, in resistance movements in Europe, code words conveyed information about operations or people. Even today, the sound of the first four notes from Beethoven's Symphony no. 5 cause a tingling in the spine for many who lived through World War II, especially, but not only, in Europe.

REACTIONS TO DEATH AND THREATS OF DESTRUCTION

Many statements in music were composed as a consequence of World War II. Some music is purposely painful in its loudness or in its expression of anguish, as were parts of Shostakovich's symphonies, but other parts of his music, and all of some other composers' music, are meditative.

Darius Milhaud (1892–1974) fled to the United States in 1940 after the fall of France to the Germans. Friends and acquaintances had become victims of the Nazis and their sympathizers. In 1952, he wrote *Le château du feu* (*The Castle of Fire*) for choir and a chamber orchestra without violins. The words near the end are powerfully delivered. The passage "In the firmament of fire" is sung with a strong crescendo at the word "fire," followed immediately by a marked *pianissimo* as the singers sing of triumphing "over death." Milhaud was so filled with horror at what happened to his Jewish friends and to others during the war that he felt that any victory prefigured only the dawn of an uncertain life.

German composer Richard Strauss was criticized and then restricted in his movements by the Nazis (Ashley 1999, p. 188). Their arrest of his wife's family distressed him. Further, he mourned the destruction of German culture at the hand of Hitler's regime. He was also shocked by the bombing of his house in Weimar and of landscape symbols of culture (including the Dresden opera house, where all of his operas had been performed, and the opera houses in Munich and Vienna). In 1945, once free to write, Strauss, then in his late eighties, wrote *Metamorphosen* for a string orchestra, the twenty-three players each having a different part. The work opens in a dark and moody manner. Various themes are introduced, including fleeting references to works by others, most notably to Beethoven's *Funeral March*, which is used in the final ten bars to quietly close the work. The symbolization of the death of culture is profound.

The death of culture or the death of people? Which is more important? For Milhaud, it was people, but he feared the future due to what some humans had done to other humans during the war. In contrast, for Strauss, a vain man, culture won out. He did not write about the consequences of war for people. Other composers give priority to their anger and horror over the deaths of millions of people. Many approaches were taken, from a focus on one person to many. British composer Bernard Stevens composed most of *A Symphony of Liberation* during the blitz on London and completed the work in 1945. While dealing with the British people's situation in general, it is dedicated to the memory of his friend, painter and poet Clive Branson, who was killed in action. Regrettably, I know only the second movement, "Resistance," a scherzo, which vividly conveys a feeling of people's strength and courage in the face of terrible adversity.

TWO YOUNG GEOGRAPHERS, BOMBING, AND MUSIC

For a very personal perspective we can consider the writings of the geographer Harm de Blij (2000), who has identified how, when he was a young boy, his family's life was extremely difficult in the Netherlands during World War II. Out of that awful experience, he decided to become a geographer, which he succeeded in doing, and he is now a leading professor and geographical commentator. He is also a violinist, a student of his musician-conductor father. While witnessing the consequences of Allied bombing and the presence of German troops, a psychological survival technique for his family and friends was the regular gathering at his family's house of adults who performed chamber music together. To everyone's shock, a German military officer, clearly hungry for the joy of music, appeared one day and joined in, listened to the chamber music performance, and then just as abruptly departed. Though his personal action was against German military code, he could nevertheless have turned in the de Blij family and friends for congregating in an "illegal" manner, but he did not. Such is the power of music.

In my own case, I have memories of listening to music from the radio in our house in Glasgow, Scotland. Johann Sebastian Bach's *Jesu, Joy of Man's Desiring* lingers in my mind as the earliest piece of music I heard. Music apart, my wartime experiences included a brick bomb shelter at the bottom of our garden. After several air raids, however, my mother simply had the three children go under the kitchen table whenever the sirens went off, her theory being that if a bomb hit that close either to the shelter or the house, we would all be wiped out anyway. Our bomb shelter thereafter became the best-made hen house in Glasgow! Any thought of an idyllic childhood was crushed by the terrible memory I carry to this day of a house three from ours receiving a direct bomb hit which killed everyone there; yet, the memories of music from the radio also remain, and in contrast, they are sweet. As with de Blij's experiences, my story is that of a child, not an adult. We were both fortunate, for countless others suffered terribly before and during the war, and many died or were killed.

THE CAMPS

French composer Olivier Messiaen (1908–1992) was a prisoner of war in Stalag VIIIA in Silesia from 1940 to 1942. He maintained his psychological well-being by composing. *Quartet for the End of Time* is a haunting, seven-movement work, the last movement of which relates how, in a tangle of rainbows, an angel announces the end of time. The movement concludes with a grand climax that represents the ascension of the Son of God to His Father, and of humans to God. The unusual instrumentation—clarinet, cello, piano, and violin—was chosen because these were the only instruments available in the camp. The Germans permitted the work to be premiered in the camp; five thousand fully attentive prisoners recognized the work as an expression of hope in their time of deep despair. Messiaen was fortunate. His life was spared. Others were less fortunate. After the war, Messiaen wrote *Et exspecto resurrectionem mortuorum* (*And I Await the Resurrection of the Dead*) (1964) for wind, brass, and percussion. Each of its five sections is headed by a biblical quotation. The work opens with a feeling of profound despair but moves gradually, with an air of mystery, to a feeling of joyous exaltation. Using a rich array of colors, silences, and dramatic bursts of sound, the work challenges the mind and the soul.

Mourning, praise, peace. These are powerful issues. How does one deal with them when the subject concerns specifically those who died during the Holocaust? A different approach to remembering those killed is to be surprised by their joy, their expressions of hope, and their recognition of the dreadful reality that they were doomed. Many artists of amazing talent were among the hundreds of thousands of Jews annihilated during

World War II. Around them in the concentration camps were people coming in from afar by train, being sorted, many exterminated in gas chambers or hanged or shot, while a few were spared, only to be sent elsewhere to continuing uncertainty (Hewitt 1997b). The horribly dehumanizing landscape—barbed wire fences, watch towers, rows of bleak barracks; crowds of people with heads shaved, in shabby clothing, always being counted, with inadequate food, and just outside the fence, the German officers' mess and soldiers' barracks with obvious benefits denied the prisoners; the sight and sound of trains carrying the helpless, people being sorted with those in this line or that, soon to disappear forever; the piles of clothes and jewelry taken from prisoners; the stench from the gas chambers; perhaps a factory or a quarry in which some prisoners were forced to work, always under the watchful eye of armed guards—could so easily have inspired only torment and despair. Yet, in some camps, notably Terezin, several composers—and painters and writers—managed to remain creative. An orchestra was formed, and it played music by composers in the camp. The musicians were never sure who would be present in the camp the next day to perform any score; hence, parts often had to be rewritten to accommodate whoever was still alive. What kept the composers and musicians going? Were they inspired by a need to express their inner selves, their souls, in the face of darkness? Or could it have been that by composing, they were able to escape in their minds, for a moment or two, from the tyranny of the landscape of evil in which they found themselves? We will never know. We do know that works created in concentration camps can still inspire, for many of the orchestral works survived, even though those who had created the music were slaughtered by the Nazis.

Robert Evans's cantata *For the Children: No Silence of the Soul* (1997), for two narrators, children's chorus, and chamber orchestra, captures the poignancy of poems written by children aged eight to fifteen in Terezin. Most of the adults and children imprisoned there were eventually sent to nearby Auschwitz-Birkenau to be put to death. Evans notes that the prisoners "could have been defeated by the daily suffering and oppression to which they were subjected, but instead took every opportunity to liberate themselves from it" by creating a cultural life. Few of the children survived, whereas more than four thousand drawings and hundreds of poems written by them were hidden from the camp guards and have survived. The poems do not focus just on uncertainty and impending death, though these are evident, but instead they often mention beauty that once was in their lives, including memories of flowers and butterflies. Showing remarkable insight, the children juxtaposed lines about happy times in Praha, Budapest, and other places with biting comments on camp life (the lack of bread, the watery soup, the disappearance of a friend) or on what being in the camp meant about the lack of a future. Using a selection of the poems and drawing from his talks with some Holocaust survivors,

Evans composed *For the Children*. The poignant geography of the mind thus created in song and orchestral sound stands as a remarkable memorial to the lives lost. It also reminds us of the strength of the human spirit. Evans has also written a tender four-part song, *Shalom*, in remembrance of the March 1996 massacre of the sixteen children and their teacher in Dunblane, Scotland. Part of Evans's comments on Terezin also apply to Dunblane. He wrote, "Life is real, not virtual; children are real and precious, as witnessed by the stunning wonder of these poems. I only hope that someone will listen and that I have, with my music, done them justice. It should be a bittersweet remembering or a nudge not to be forgotten."

Millions of Jews, Roma, gays, people with mental and physical disabilities, and Christians of conscience and other objectors to Nazism died at the hands of the Nazis. Many works have been written in commemorative fashion, but none may be as pungent in its expression of horror, pain, and anguish than Arnold Schoenberg's 1947 composition, *A Survivor from Warsaw*. A cantata for narrator, men's chorus, and orchestra, this stunningly compressed, single-movement work lasts for about six minutes. So taut is the piece that the listener feels the seams will burst. Expressed in Schoenberg's twelve-tone technique, the work contains irregular rhythms and, at times, is jarring to the ear. The narrator is a Jew in a Nazi concentration camp, speaking from the prisoner's perspective. Nazis shout orders in German to some old and sick men, who are beaten, then forced to line up and be counted prior to being led to the gas chamber. This emotionally draining scene suddenly takes a sharp turn when, as a musical crescendo reaches its peak, the narrator's voice is unexpectedly overcome by a stunning expression of defiance. The men, with a trombone soloist, strongly to a steady pulse, as if to underscore the certainty of their faith and their resistance, sing a traditional Hebrew prayer, "Hear, O Israel." This song of supplication concludes the work. The music in the work has harsh tensions, until, at the end, there is that strong steady pulse. Opening with ghoulish, instinctual, brutal harshness, the work ends on a sublime level. Is Schoenberg relating hell and heaven on earth? Certainly, hearing this powerfully sensitive, but deeply disturbing, work on how survivors must have felt—with its emotional extremes, distorted images, high tension, and the awesome moment when the narrator's message is overcome by the power of a crescendo that leads to the certainty of the prayer—is like being kicked in the stomach. How can one applaud?

THE BOMB

Thousands of Japanese civilians died as a result of the dropping of atomic bombs on two of their cities in August 1945. Polish composer Krzysztof Penderecki (1933–) wrote *Threnody: To the Victims of Hiroshima*, first performed in 1960. Just as the bombs were like nothing else before, so, too, was Penderecki's music. Fifty-two string players have to use nontraditional techniques; they must strike their instruments to get a percussive effect, they must count both playing and between-playing durations in blocks of time marked in seconds on the score, and they create no melody. Entrances are hard for a listener to anticipate. The sounds seem at odds as violins play high, piercing pitches, while the cellos and basses growl out very low pitches. The instruments then play in unison only to flare off once again into conflictual sounds. The music builds toward a horrifying climax. No resolution is expressed; instead, a long steady diminuendo starts and eventually reaches silence. As with Schoenberg's *A Survivor from Warsaw*, Penderecki's *Threnody*, an angry expressionistic work, grabs the listener by the throat. In its totality, the work is shocking, an affront to the senses and to the soul. Both works are emotionally draining for any audience (and the musicians). Soundscapes of shock and dismay are created, as if to say, we must not forget.

INNER STRUGGLE AND MOVING BEYOND PAIN

Shostakovich faced hardships and rewards because of his prominence as a composer in the Soviet Union. He was limited, it is thought, in how far he could go in revealing his inner personal feelings. Another notable composer, Gustav Mahler, had no such limits placed on him by a political regime; yet, he struggled with many battles, some internal (a constant searching for meaning in his life, as discussed in chapter 8), others external (overcoming bigotry aimed at him by prominent members of Viennese society, including some of his professors). He revealed his agonies, searches, findings, and remaining questions in his work.

Another composer concerned with the meaning of life and the hereafter, Richard Strauss (see above), wrote an apocalyptic score in 1889, when he was twenty-five. The dramatic tone poem *Tod und Verklärung* (*Death and Transfiguration*, op. 24) tells the story of a sick man close to death. He suffers greatly from pain, which at times tortures him. After one attack, he recalls his life, thinking about his childhood and about when he was a youth filled with desire and passion. His pain returns as the goal of his life journey appears, that is, his ideal for reaching perfection in his art, an ideal that is impossible to reach. He dies. His soul leaves his body, whereupon he discovers in the "eternal cosmos" (Strauss's own words) the realization of his ideal. Strauss also wrote *Ein Heldenleben* (A Hero's Life, which was actually about himself), a set of five tone poems, one of which, "The Heroes Battlefield" (part four), has a distant fanfare of trumpets (offstage), and then the orchestra becomes the battlefield. After some remarkable sound pictures, a triumphant orchestra evocation signals the hero's victory. This is followed by "The Hero's Works of Peace" (part five), which celebrates peace and his spiritual development.

Composers sometimes are triggered to write following the death of a loved one. We all die, but not always easily. Many people must fight a "war" with cancer, Alzheimer's disease, or some other illness. Others are struck by heart attacks or are hit by vehicles. Some are murdered. Whatever the cause, even if there is mass death, as when a building collapses or bombs fall, death comes to one person and to another. There may well be collective grief, but individual grief is most poignant. Some people may get stuck in the mire of grief, while others are able to mourn and use that experience as the springboard to new thinking, perhaps with new insight. Ellen Zwilich, for instance, was married for only nine years when her husband died suddenly. At the time of his death, she was composing her *Chamber Symphony*. It was a personal struggle for her to return to composing, but when she did, she realized she had changed, as had the purpose of the composition. In its final form, written in Joseph Zwilich's memory, her work remembers "beauty, joy, nobility and love—greater realities which artists must learn to express again" (*New York Times*, July 14, 1985, quoted in Jezic 1988, p. 176). Her message, it would seem, is that mourning is okay but that one must go beyond it. Robert Evans agrees: he responded to the death of his father, younger brother, and two friends by composing *Pie Jesu*, which helped him to mourn and then to move on. And, as noted earlier (p. 110), Gillian Whitehead wrote *Ahotu* (*O Matenga*) for her dying father in response to his request for sustenance as his spirit traveled "home."

MOURNING, RECONCILIATION, AND HOPE

People react in different ways to trauma and death. Once war is over, why not simply forget about it? Many people tried to do that after World War II, through denial, as did many Germans and Japanese writers of school textbooks. Others acted to construct something new, something that would serve to bind people together rather than call up old allegiances and enmities. It is rare to be able to forget completely, thus memorials abound on the landscapes that have been the sites of conflicts. They also exist on the landscapes of countries that sent soldiers, nurses, and others to fight in far away places. At least one day is annually set aside in many countries to officially remember and celebrate the brave actions of members of previous generations lost in conflict. However, some people cannot forget; their suffering continues, long after a conflict. Mourning—which implies the emotional passage from one state of mind to another—is needed.

Symphony of Sorrowful Songs, Symphony no. 3 (1976) by Polish composer Henryk Górecki (1933–), uses texts from several sources about three mothers and the deaths of their children. In the first, a fifteenth-century lamentation found in a monastery, Mary mourns the death of Jesus. In the second, written in Górecki's lifetime, a mother scribbled a prayer for her child on a prison-cell wall during the time she was suffering violence at the hand of the Gestapo. In the third, drawn from a folk song, a woman's child has just been killed, this one a soldier killed in battle. The soundscape is poignant. It is intense in its mourning. Very slow moving, it conveys what some have called "a metaphysical calm; though others experience only maddening stasis" (Steinberg 1995, p. 173). For some, the music is haunting, as death is mourned. One does not get a sense of hope from this work; it, rather, lets one ruminate on losses.

Leonard Bernstein wrote Symphony no. 3, *Kaddish* (1963), a powerful "prayer" for orchestra, choir, a soloist, and a speaker. The word "kaddish" is Hebrew for "sanctification." The prayer is offered upon a death, at a burial, on memorial occasions, and, indeed, at all synagogue services. Intriguingly, whereas it uses the word *chayim* (life) three times, no mention is made of death. In that sense, then, it is not a prayer of mourning as much as of remembrance. As one hears Bernstein's *Kaddish*, it is impossible not to reflect on those who died during the Holocaust caused by the Nazis, supported by virulent anti-Semitism among people in many European states. Bernstein wrote the work as a form of remembrance but also as a statement about the importance of life. Significantly, praise and peace are at the core of the prayer.

FUNERAL MARCHES AND REQUIA

Frédéric Chopin's *Funeral March* was mentioned at the outset of this chapter. Many additional funeral marches have been composed, some pertaining to war, others not. As noted earlier, Gustav Mahler includes a funeral march in Symphony no. 2 (as he did, as noted earlier, in most of his symphonies). The adagio (slow movement), with Brucknerian fullness and harmony, is a grand hymn to humanity's salvation through spirituality. French composer Lili Boulanger (1893–1918) wrote *Pour les funérailles d'un soldat* (1913) for baritone, chorus, and orchestra, and during World War I, she wrote *Marche funèbre* (1916), but this piece has since been lost. Perhaps the grandest funeral march of all is the *Symphonie funèbre et triomphale* (1840) by Hector Berlioz. The work is constructed as follows: funeral march, funeral oration, apotheosis. It calls for huge forces; Berlioz used six hundred musicians (most of whom were soldiers) during the premier performance. The act of performing *Symphonie funèbre et triomphale* created a distinctive and colorful military landscape in Paris that day. However, to Berlioz's dismay, the military officers were less than successful in keeping control of the troops.

The compositions of remembrance by Bernstein and Evans, discussed above, are not requia. To be such, a work, theoretically, if not in fact, must follow the standard rite of remembrance for the dead of Christendom, as in the Roman Catholic, Anglican, and Eastern Orthodox traditions. *A German Requiem* by Johannes Brahms is misnamed in the sense that it is not technically a requiem; yet, it stands as one of the most beautiful works of remembrance in music. There are, of course, hundreds of requia, some named for important military persons, notably the beautiful *Lord Nelson Mass* by Franz Joseph Haydn. Formally known as the *Mass in Time of Anguish* (*Missa in angustiis*), it was written in 1798 and performed in Admiral Horatio Nelson's presence upon his return to England. The "Kyrie," it is said, expresses the despair of the Austrian nation as Napoleon's armies swept across Europe. And the trumpet and timpani calls that punctuate the "Benedictus" represent the battle in which Nelson defeated Napoleon's fleet in the battle of the Nile (Marta McCarthy, personal communication, December 4, 2004).

Berlioz's *Grand messe des morts* (1837) or, simply, *Requiem Mass*, was composed to celebrate the sacrifice of the "more or less heroic victims" of the July 1830 French Revolution. In "Dies irae" and "Tuba mirum" the brass in the orchestra, plus the four brass bands located in disparate parts of the performance space, are used to summon the Day of Judgment with an earsplitting blast, which, when coupled with thunderous chords from sixteen timpani, give voice to one of the most remarkable sounds in all of music.

Benjamin Britten was troubled by the apparent fact that the act of living can destroy life's highest quality, its innocence, especially the passion and innocence of youth. Michael Tippett holds that Britten's "pacifism and his feeling about international war sprang from something of the same source" (quoted in Bowen 1995, p. 71). As an avowed pacifist, Britten wrote a biting satire, *Our Hunting Fathers* (1936) in response to what he saw happening in Germany under the Nazis and in anticipation of a war. Of note, Ralph Vaughan Williams also wrote a work against war, his Symphony no. 4 (1935), but neither work had any impact on political or military decision makers. A former colleague, David Gardner, suggested to me that these two works form part of the corpus of artwork that stands as "the conscience of survivors." He placed the work of Vaughan Williams and Britten alongside the powerful paintings by Francisco Jose de Goya (*The Third of May*) and Pablo Picasso (*Guernica*). I can attest to the acceptance of these shockingly vivid paintings as gruesome indictments of human cruelty.

Benjamin Britten's *War Requiem* (1962) is one of the most powerful works to flow from the end of the World War II. As totalitarianism spread and war started, he wrote a precursor, *Sinfonia da Requiem* (1940). After the war's conclusion, he felt it was right and necessary to stop and mourn the dead, but he believed foes in war should do it together in the hope of achieving reconciliation. He expected thereby that the perceived differences that had led originally to war could be set aside so that there would be no more war. Britten's *War Requiem* thus goes beyond the landscape of horror conveyed by Shostakovich or the soundscape of mourning created by Górecki and, instead, ultimately creates a landscape of reconciliation and hope. The requiem, in remarkable fashion, adjoins the Latin text of the Mass for the Dead with poems in English by Wilfred Owen, a young Englishman who was killed in France just one week prior to the armistice that ended World War I. The poetry was written when Owen was a young officer in the trenches. His poetry is honest, without bravado or false heroism. For example, he wrote, "What passing bells for these who die as cattle? Only the monstrous anger of the guns." And, with reference to the use of mustard gas, he wrote, "Out there we walked quite friendly up to death. . . . We've sniffed the thick green odor of this breath." In "Dies irae," the tenor and baritone sing interwoven texts, expressing a living grief for the dead; at the same time, the dead describe their horror to the living. The tragedy of lost lives is caught in the words that men die not for men but "for flags." This section ends with stunning trumpets of the Last Judgment evolving into the sad bugle call to battle. Later, at the end of "Benedictus," "Hosannas" are sung with a brilliance that stuns the mind. As this section closes, there is a profound sense of despair, but Britten suddenly transforms that feeling into one of hope by means of an unexpected change of key. The emotional switch is breathtaking. The "Libera me" is intense, as the full orchestra overwhelms the choir's reiteration of the "Dies irae," and the work then has the two protagonists singing "Let us sleep now . . . ," as the choir sings with the soprano soloist in a beautiful descant. The juxtaposition of all of the forces is amazing, and frightening, perhaps most notably at the moment when the baritone sings, "I am the enemy you killed, my friend." The work concludes with a quiet "Requiescat in pace" (may you rest in peace).

War Requiem requires two clusters of performers. The Latin text is delivered by a mixed (male-female) chorus, a boy's choir, and solo soprano, with a full symphony orchestra. The poetry, in English, is sung by tenor and baritone soloists and a separate chamber orchestra. The orchestras and their respective singers play against each other at times in an intriguing way, almost as if one is trying to find the other, while at other times one is echoing the other. The musical styles used vary from fugue, to aria, to Gregorian chant, in a manner that is at once simple and direct and yet most challenging to both the intellect and the emotions. Significantly, the soloists for the premier performance were Russian (Galina Vishnevskaya, soprano), German (Dietrich Fischer-Dieskau, baritone) and English (Peter Pears, tenor). There is no rejoicing in this work. The English text conveys, in John Culshaw's (1963, p. 5) words, Owen's "intensely personal vision of a man driven to an extremity of action and emotion—the extremity of war, and the grief of man for man." With more than four hundred performers, the first performance of the work was given in May 1962 in the new Coventry Cathedral that was constructed beside the bombed debris of the old cathedral—new life out of the destroyed. The emotional response to that performance was powerful, and so it is whenever the work is performed. Significantly, Britten (1962, cover page) prefaced his work with a few telling words by Owen: "My subject is War, and the pity of War./The Poetry is in the pity . . ./All a poet can do today is warn."

HEAVEN AND ELSEWHERE

Until it is "too late," as it were, we cannot know what music is played in heaven. However, I can report that from tombstone epitaphs and children's paintings, I "know" (Knight 1986) that heaven is "up there" on top of clouds! Further, as an aside, based on cemetery-name research in the United States and Canada by Wilbur Zelinsky (United States), a leading cultural geographer, and Dennis Taylor (Canada), one of my students, it is possible to add that there are varying types of landscapes in the afterworld, including, at least, distinct Canadian and U.S. neighborhoods (Zelinsky 1976; Knight 1986).

What about music in heaven? Four hundred years ago, Friar Martin of Cochem wrote that surely the saints "will play music together and sing psalms," while "the choir of archangels [will pay] homage" by "singing as praise [and] for enjoyment" (McDannell and Lang 1988, pp. 126, 172–75, 215–16; see also Russell 1997). Perhaps that is all we need to know. On the other hand, geographer Yi-Fu Tuan (1998, p. 175) suggests that we must surely anticipate the sound of the clichéd harp, but "to a music lover, the thought of a Bach fugue or a Mozart sonata in heaven is not out of the question." And the English writer Neville Cardus (quoted in Watson 1994, p. 229) wrote of the beginning of Mozart's Piano Concerto no. 23 in A major, "If any of us were to die and then wake hearing it we should know at once that (after all) we had got to the right place."

There is, of course, other music that one might expect or hope for, but here we get to the matter of personal preference. I for one would hope to meet Anton Bruckner (1824–1896) to say thanks for his magnificent spiritual soundscapes and then to sit down with him to listen to any of his works, but especially his Symphonies nos. 8 and 9 and *Te Deum*, for they surely were inspired "by heaven." British composer Robert Simpson (1921–2001) would already be talking with Bruckner, so I would be happy to meet him, too, to say thanks, especially for his (Simpson's) Symphony no. 9, and to see if Bruckner approved of what Simpson had written about him and his music (Simpson 1991). But what if there is no heaven? What is the alternative? Clues come to us from several composers; their respective hereafters are on an island, "over there," or "down there."

If there is no heaven in the manner described above, then, if some composers are correct, there is only an island, somewhere, to which people are taken. (Others would suggest that the island is actually purgatory, or "the waiting room," or whatever other euphemism is used.) A number of compositions represent islands of the dead, including *Die Toteninsel* (*The Isle of the Dead*) by Max Reger (1873–1916). Composed in 1913, it is one of four symphonic poems based on paintings by Arnold Böcklin. The scene is haunting, with rich, romantic sound, conveying a powerful feeling of tranquility. Jean Sibelius conveys another famous isle in beautiful musical language. Inspired by Tapio, the ancient god of the forest, Sibelius's composition remarkably recounts the story of a swan swimming on the waters that circle Tuonela, the land of the dead, or hell, in Finnish legends. The *Lemminkäinen Suite* (*Four Legends*) comprises four tone poems, the second two of which convey a strong sense of darkness and gloom. Of the four pieces, the third, *The Swan of Tuonela* (1893), is best known, and, arguably it is the masterpiece in Sibelius's repertoire. A mournful, haunting, and yet truly lovely melody is sung by the cor anglais with somber strings and low wind harmonies in the background. The cor anglais represents the swan gliding endlessly around in the dark waters. One cannot escape death, it seems to sing, adding that the life hereafter is both bleak and unending.

The Ride of the Valkyries is a less somber work that represents the exultant ride of the warrior maidens through angry skies as they carry the bodies of dead heroes to Valhalla, the castle home of Wonton, the king of the gods. This work opens the third act of Richard Wagner's opera *Die Walküre* (1870). It is one of the best known of all of Wagner' compositions and lasts a mere six minutes. In contrast, performances of the three-opera cycle, known as the *Ring of the Nibelung*, last for nineteen and one-half hours. To some, these operas are wonderful to experience; to others, the experience is hell itself!

Yet another expression of the hereafter is depicted by Tchaikovsky in what may be his finest composition of program music, namely, *Francesca da Rimini: Fantasia after Dante* (1876). The work is continuous, with three sections. In the middle section, a solo clarinet introduces a beautiful love melody representing Francesca and her ill-fated love story. The two outside sections relay a frightening vision of the inferno. The music vividly renders images of terror-filled lost souls, first by dramatic, misery-laden chords and then, as momentum gathers, by a terrible whirling tumult. Tchaikovsky's hell is not a calm island but a fearsome place.

American composer Steve Reich's *The Desert Music* (1984) reminds us that total hell on earth could be triggered by human ignorance, suspicion, fear, and stupidity. The work raises the tragic promise and consequences of a human-made holocaust. The music evokes the atomic bomb testing in the U.S. Southwest from the 1950s on and reveals the scarred emptiness of the earth after a nuclear blast. The rhythmic feel of time passing, coupled with a lean orchestral sound, creates a feeling of time too late and a creepy sense of nonbeing. *The Desert Music* is about what the earth could become, unless suspicions, we-they tensions, hatred, fear, and nuclear proliferation are dealt with in new ways.

Earlier in this chapter, "Mars, the Bringer of War" was identified, so it is only right also to identify the second movement in Holst's *The Planets*, "Venus, the Bringer of Peace," which may appropriately symbolize the anticipated peace of heaven. "Venus" opens with a gentle, four-note horn solo, joined by the woodwinds, repeated, and then developed in a series of variations by the remainder of the orchestra in a slow-moving, gentle manner. Peace is indeed the emotion one develops from hearing this work, although, of course, the listener would have just heard "Mars" in a concert program. The let down after hearing the highly charged, jagged edged, and aggressive "Mars" is therefore pleasant, to say the least. Listening to "Venus" has a cathartic effect.

MUSIC'S CATHARTIC ROLE

The performance in Leningrad in August 1942 of Shostakovich's Symphony no. 7, the *Leningrad*, proved cathartic for the survivors in that still-beleaguered city. The audience wept and cheered. They were living through an atrocious time, and, here, their struggle was expressed in music by one of their own, who had been with them in the early days of the city's soul-destroying siege by German forces.

Another instance of identification with a people suffering and a form of catharsis occurred in Orchestra Hall, Chicago, in January 1969. In August 1968, Russian tanks and troops had smashed their way into Praha, Czechoslovakia, to crush the democracy movement. In memory of those who had lost their lives at the hand of the invaders and to mourn the setback to those fighting for freedom and democracy, the Chicago Symphony Orchestra, led by Czech conductor Raphael Kubelik, performed Bedřich Smetana's full cycle *Má Vlast* (see chapter 5). Tension in the building was high, the playing intense. When the final note ended, silence held the air as if people dared not breathe, but, then, release came, and the audience felt free to shout and cheer and clap for a very long time. Tears were abundant, as people thought of the violence still underway in Czechoslovakia. The emotion of the moment, generated by the powerfully evocative score, touched musicians and audience alike. After fifteen or so minutes of tribute, Kubelik led the orchestra offstage; otherwise, the players might have waited onstage for half an hour or more before the emotional outpouring stopped.

Music can also have a personal, cathartic effect for a composer, as noted earlier for Zwilich and Evans. When Antonín Dvořák died in 1904, followed by his dear Otilie a year later, their son-in-law Josef Suk (1874–1935) was moved to write a five-movement symphony in their memory. The *Asrael Symphony* "is saturated with a sense of loss, conveyed in music of cathartic intensity," and continuing, "in the depths of his grief, Suk wrote that he was 'saved by music'" (Staines et al. 1998, pp. 506–7).

Finally, some years ago, a neighbor's teenage child hanged herself. In grief, and with memories of my sister-in-law, who also committed suicide, I listened to a recording of Carl Nielsen's Symphony no. 4, the *Inextinguishable* (1916). Nielsen said of "the elemental Will of Life," "Music is life and, like it, inextinguishable." The work is restless and driven, being full of tension as competing forces contend with each other. This tension culminates in a battle between two timpanists (at opposite sides of the stage), who thunder out massive expressions of fury at each other. In the end, however, the hurtful tension is released by the return of a sweet melody from the first movement to signify a remarkable resolution, namely, that the human spirit will survive. That battle "landscape" surely described some of my inner feelings, and the resolution provided by Nielsen helped me reach a new understanding of what was possible. By listening to that work over and over again for three or four days, I worked something out concerning the two deaths, and I was thereafter better able to handle my feelings. Yes, music can indeed speak for and to the spirit! Such may not be cultural geography, but it is real.

BITTER REMINDERS

There is a need to remember that many people in the world suffer a lack of freedom of thought and action. Many of China's leading composers of today suffered during the so-called Cultural Revolution that ended in 1976. A number of them have been able to write music related to that terrible time, for themselves and thousands of others. Wang Xilin's music relentlessly represents the cruelty experienced during the "revolution" by all intellectuals, not just composers for, he says, "I want to tell people about the tragedy and darkness of Chinese history. This is my music" (Melvin and Cai 2001, p.30). He had graduated from the conservatory of music just in time for the revolution, only to be exiled to Shanxi Province for fourteen years because he had dared to criticize publicly Mao's declaration, akin to that by Stalin, that "art must serve politics." His experiences during those fourteen years were appalling. Another Chinese composer, Sheng Bright (1955–), composed *H'un* [*Lacerations*]: *In memoriam 1966–76* (1988), a mournful, dissonant, orchestral evocation of the suffering of the Chinese people during the Cultural Revolution. In 1990, Chou Wen-chung wrote *Windswept Peaks*, which, he said, was "composed during the democracy movement that culminated in Tiananmen Square, [and it] is dedicated to persecuted Chinese intellectuals, who 'stand tall among the mightiest peaks in the history of humanity'" (Lee 2001, p. 789).

THE COMMON PERSON?

People detained or arrested for doing something the state has disliked are not necessarily special people. Most such people are "ordinary" in the way they carry out their lives. In a sense, they are made special by the act of the state. Whenever a state, perhaps as a police state, goes after some of its people, the oppressed may carry out heroic counterreactions; however, there is no guarantee that such will occur. Further, everyday events can cause stress and strife in the lives of ordinary people, no matter where they live. Prompted by intense anger and frustration at having his apartment burgled, Canadian composer John Estacio (1966–) wrote *Victims of Us All* (1995). The work goes far beyond the initial stimulus for it mourns all that we in modern society suffer from due to the violence of the age. His riveting landscape of anger, frustration, and reestablished cathartic calm is a reminder of every person's ordinary, daily trials, which, one by one by one, combine to represent aspects of the social landscape around us, of which we are but a part, and of our common need to find a sense of peace.

CONCLUSION

Numerous composers have demonstrated that music has the power to speak of unpleasant things, of survival and remembrance, and also of reconciliation and hope. What will composers write in the future about reconciliation and hope? Can composers suggest new ways of thinking as they reflect on the present? Will they build on the past or create new musical forms and, thus, new landscapes in music? Will the day ever come when there will be no more landscapes of war and destruction? At the risk of sounding trite, only time will tell. Meanwhile, humankind will be able to benefit from the sometimes disturbing insights, yet comfort, to be gained from the questioning and presentation of tentative answers by countless composers who provide us with music for orchestral performance. Geographers are "in business" because they examine differences in time and space. Composers do also, so perhaps new programmatic orchestral music on some of the current landscapes of suffering will be forthcoming. However, it is to be hoped that in composing their music, composers will be able to include expressions of hope, as Robert Evans does in his composition on drug rehabilitation among youth, entitled *Bridges* (2000) for choir and orchestra. Life without hope would trap us all in a landscape of suffering and despair. The work best associated with hope may well be the single most identified theme in classical music, namely, *Ode to Joy* in Beethoven's Symphony no. 9. What can be said about this astounding work? Perhaps it enough to repeat something a musician colleague said to me at the end of a performance of the work: "Heaven has just spoken."

8

Music in Places

We now turn to places where music is performed, some indoors, others outdoors. As noted earlier, geographers are concerned with space and its organization, and such are our concerns here. This chapter presents some summary comments on the history of performance spaces; identifies some buildings constructed for the performance of music and some aspects of the spatial organization of orchestras and spatial music; remarks on the interplay between nature's soundscapes and music performance; and comments on special localities for festivals, the use of stadia, annual tours, the role of radio and television, and the use of visual aids.

PERFORMANCE SPACES

Most cultures celebrate the links between humans and "mother earth." For some, music has a divine nature that, in its most complete and lucid expression, reflects nothing less than the perfect harmony of the whole universe (Khan 1996). Composers have long addressed the link between humans and the divine, and some, as noted earlier, believe themselves simply to be the conduits for music derived from "on high" so that it can reach paper and thereafter be performed. And some people attend worship services because of the music rather than the spoken word; for them, the ministry of music is crucial.

A full discussion of music in worship is beyond the scope of this volume, but that the performance of works pertaining to worship can speak to the soul and to the intellect must be acknowledged. Hearing Serge Rachmaninov's *Vespers* in a former Leningrad (St. Petersburg) church stripped by Soviet authorities of its religious trappings but kept as a performance building stands as one of the most remarkable experiences in music I have had. The Leningrad (now, again, St. Petersburg) Chamber Choir, singing a capella (without an accompaniment), was superb, the dynamics of the sound in the huge bare space stunning, and the total experience extremely moving. For many, hearing Bach's *Mass in B Minor* or perhaps Mozart's *Coronation Mass* can be like no other experience. As a reflection of beliefs, these works can "speak" to people beyond religious structures or constraints and, so, can have a powerful impact, even when played, as they often are, in the concert hall. However, religious music performed within religious spaces can take on an added dimension that is hard to replicate in a concert hall. The presence of the divine, in sanctified space, may have meaning far beyond what a skeptic is able to deal with, and the inner message and power of sacred music, even when performed in such space but apart from a service, may speak to the audience in ways that are hard to explain. It is when heard in a church or cathedral, where the music is associated with the liturgy, with its structure and content, that the music makes full sense.

Early Christian church spaces were originally houses, sometimes with walls knocked down between rooms and even between linked structures to obtain larger areas for worship. In time, special buildings were constructed. A key part of such structures was the "choir," a word derived from Greek *khorus* for the section within a church where the singers and dancers were located. Dance and music became separated, with the term *choir* losing any association with dancing. Importantly, the choir, as space, is located between the nave (where the commoners sit) and the altar presbytery (the most sacred space within the overall sacred place). The choir, sitting in the choir, or chancel, thus serves as the intermediary between the people and the priests, whom, in turn, some perceive to be the intermediaries between them and God. Interestingly, the Greek *orkhēstra* was derived from *orkheisthai* (to dance) (Whone 1990, p. 45). Places of performance were thus initially places of dance *and* music. Ancient Greek theaters were outdoors, with rows of seating areas on a hillside looking down into the amphitheater where the performers were situated. Later, when performers were in palaces, they generally performed at the same level as the audience, for, among other things, they could not be seated higher than the ruler. With time and wisdom came the idea of elevating the performers above the audience level; hence, stages were constructed.

When, eventually, players were no longer identified with the performance space, the term *orchestra* became associated only with the players, not also the space. Interestingly, however, the word "orchestra," still refers to space, namely, to the ground-floor audience area. Over many centuries, performance spaces evolved into huge halls with balconies around a higher level (or levels) and all seated in the "house" being able to see and hear the performers on the stage. The idea of having large buildings just for the performance of music climaxed during the nineteenth century as civic governments built grand structures to house "their" orchestras. A notable feature in such orchestral halls was often a splendid organ, with grandiose pipes. This tradition continues.

There is no denying that a great performance in a great performance place can be exciting and may evoke a visceral response that is both puzzling and wonderful. Clearly, a structure wherein performances take place can be important in this process of hearing, but perhaps, ultimately, the listener's mind-set is more important. A building, be it a church or hall, normally is constructed to represent physically, if not also symbolically, a separateness from the outside world. Cathedrals and concert halls are *closed* containers: they are for temporary habitation. The audience becomes involved in directed, or focused, listening within the structure. The performers "speak" to the audience, who normally sit facing the platform upon which the performers are located. If the acoustics are good, the audience will feel as if it is in the middle of the sound, which can result in a wondrous feeling for the audience. The sounds cannot escape the container; they remain inside.

MORE THAN A BUILDING

Any orchestra is obviously more than the building in which it performs, but the "hall" is vital if an orchestra is to do well, in terms not just of excellent acoustics but also of access. Thousands of amateur and semiprofessional orchestras do not have their own space and, thus, have to use school, church, civic, or university facilities for rehearsals and performances. Many orchestras, including major symphonies, are long-term tenants in their principal performance buildings, though some own their structures. Numerous cities, especially those with younger orchestras with limited endowments, lack such a facility, much to the frustration of artistic directors and musicians. Many major European and U.S. orchestras have their own halls. European orchestras rely heavily on the state (and local government) for financial support. Such tends not to be the case in North America, where orchestras are heavily reliant on private money, though some state (or provincial) and local governments in North America provide limited financial backing for "their" orchestras, the amount varying from place to place. The two exceptions are the federally funded Kennedy Center in Washington, D.C., in the United States and the National Arts Center in Ottawa, Canada.

The buildings in which orchestras perform represent marked impresses on the landscape of the power of orchestral music. For example, the Vienna Philharmonic Orchestra has played in its magnificent Grosser Musikvereinsaal since 1879. The Boston Symphony Orchestra performed in the Music Hall from 1881 to 1900, at which time the orchestra moved into its new Symphony Hall. Many other halls are grand too, but some have not survived, including England's 1849 Liverpool Philharmonic Hall, which was gutted by fire in 1933, as was St. Andrew Hall in Glasgow, Scotland, in the 1960s. London's Queen's Hall was bombed in 1941 and the acoustically and aesthetically superior Leipzig Gewandhaus in Germany, built in 1884, was destroyed by Allied bombs during World War II. Today, many orchestras perform in buildings that remain grand landscape statements of a past era, while others perform in buildings contemporary to our times. Some magnificent buildings are not the permanent homes of any orchestra. The two best known such structures are the huge Royal Albert Hall in London, site of the annual Henry Wood (or "Promenade") concert series, and the remarkable Carnegie Hall in New York, the mecca for orchestras and other performers from around the globe. Two specially built orchestra halls help us identify several important, geographical elements.

Chicago's Orchestra Hall, constructed in 1904 as the home of the Chicago Symphony Orchestra, is "the largest, oldest, and architecturally most impressive 'cultural center' in the United States" (Mayer and Wade 1969, p. 294). Orchestra Hall, a refurbished office structure, and a new, eight-story structure were linked in the 1990s by a six-story rotunda. Together, they form a remarkable performance, music-education, community-outreach, and administrative complex. The whole complex is known as Symphony Center and stands as the city's musical center and a symbol of its cultural growth. One of the finest performing spaces in North America, the design of the hall is standard in the sense that there is a stage for the orchestra and five levels of seating: ground floor, a first balcony with boxes, and three additional levels of balconies. All audience members look ahead to the superb orchestra. Behind and above the orchestra are a pipe organ and choir balcony. A second orchestra hall, opened in 1963, replaced that built in 1882 but bombed during World War II, namely, the Philharmonie, home of the Berlin Philharmonic Orchestra. Since the designation of Berlin as the capital of the unified Germany, the structure has become the anchor of the new Kulturforum (Cultural Forum), which encompasses sixty acres of museums, galleries, and libraries. Philharmonie's three interlinked pentagons (an ancient symbol of life) symbolize harmonization between humans, music, and space (Conrad 2001, p. 2). This unusual building at once stands out because of its unconventional and bold design; yet, it forms an integrated part of the surrounding landscape due to how it was designed for the space. Inside, the generally accepted separateness of the orchestra from the audience is largely erased as the audience is located on all sides; some people are located in balconies that seem to float, while others sit adjacent to the orchestra on a low stage. One gets the sense that this is a people place, not a monumental one, since the alternating planes for seating give the impression of a human scale.

In both Chicago and Berlin, people are attracted to performances from far and wide, and not just from within the city's bounds. Each orchestra is an economic generator, inasmuch as people pay not just for their tickets but also for parking, gas (petrol), trains, meals, and other shopping while in the city centers for concerts. And, of course, the performers and the institutions spend money for goods and services in their respective communities. Each hall is an important node in the spatial organization of a city's concert-going public. It is also a node in the orchestra's system of organization, for travel away from the central facility occurs, whether just during the summer to a festival site or for a tour of the region, within the country, or even internationally. A final comment on this topic: the Berlin Philharmonie's irregular design is recognizable in the urban landscape in a way that the Chicago Symphony Center is not. The latter, at least, as represented by Orchestra Hall's Michigan Avenue facade, is but one building in a wall-to-wall series of building street fronts constructed to a set height formula within their assigned spaces as the city grew very rapidly. Both buildings, nevertheless, serve as statements that music is important. Also, neither building can be comprehended as a whole from the outside. Inside a different story applies, for, there, the focus is on the performance space. Indeed, each hall is "nothing" until someone performs. Each space, designed for orchestral performances, comes alive when performances occur.

Some music is associated with the opening of orchestral halls. The most famous, Ludwig van Beethoven's overture *Die Weihe des Hauses* (*The Consecration of the House*), was written for the opening of the Josephstädter Theater in Vienna in 1822. Ever since, the work has been played at the opening of orchestra halls around the globe. Ellen Taaffe Zwilich wrote *Celebration* (1984) for the opening of a new hall for the commissioning orchestra, the Indianapolis Symphony Orchestra. Zwilich sought to celebrate the "joyous and historic occasion with all of its inspiring symbolism of beginning and renewal" (quoted in Dyer 1986, p. 6) by drawing upon the image of celebratory bells; bells are a dominant presence in the composition.

SPATIAL ARRANGEMENTS

The spatial arrangement of players is fairly stable, based on a long-standing, standard distribution of players in the smaller chamber orchestra and larger symphony orchestra (Peyser 1986; Knight, in press). Minor changes are often made to the placement of the timpani and, sometimes, cellos and violas. Some major experimentation has upset audiences: no chairs are on the stage for R. Murray Schafer's *Cortège* (1977). Instead, the musicians march on, off, back again, and around the stage, a process repeated several times, playing as they march. For the audience, the sounds come and go, from right and left, from the rear of the stage (from behind a curtain) and the front, and so on. A listener-watcher is reminded of a college marching band strutting its stuff during a half-time show at a football game. Thus, the disjuncture between the orchestra's current performance and those that normally occurred in that space had the audience I was in either loudly applauding or jeering unsympathetically at the close of the work. The musicians, it must be said, were not altogether happy with what was demanded of them. Schafer is not the only composer to experiment with new locations for musicians, for others, likewise, have tried innovative spatial arrangements to obtain different sensory experiences (Harley 1993; Varèse 1998).

Explicitly spatial music involves the planned locations of performers throughout the hall, including the stage, as an essential factor in the composing scheme. Of this genre, Henry Brant (he was born in Montreal in 1913 but has spent most of his life in the United States) has composed more than one hundred works. His *Antiphony I*, from 1953, marks the start of the modern composing of spatial music, drawn, in Brant's case, from ideas proposed by Charles Ives. Brant's *Millennium II* (1954) has ten trumpets and ten trombones located down each side of the hall; a singer stands at the back (lobby) door, and the remainder of the orchestra is onstage. I wonder if his inspiration was some military experience, perhaps being told to "count off," with the soldiers numbering "1, 2, . . ." down the line, for the musicians do essentially that, playing one at a time, in order. One at a time, each trumpet, followed by the trombones, plays until each row is playing. Bit by bit, the audience senses that the space is filling up with sound as sound moves down the trumpet line and up the trombone line. As the singer and the orchestra join in, the audience experiences what Brant calls "sound travel" (Brant 1998, p. 238). At another time, he experimented with having the basses on the ground-floor level, cellos on the first balcony level, violas on the second balcony level, and the violins on the third balcony level to create an "entire wall space sounding at once" (Brant 1998, p. 231). *Voyage Four* (1963) is a fascinating "total antiphony," inasmuch as it calls for the musicians to be located on the stage, at the back and side walls, and under the auditorium floor. His *Homeless People* (1993) calls for a pianist onstage (playing on the strings), an accordionist in the center of the hall, and a string quartet in the four corners of the hall. In *Ice Fields* (2001), a string orchestra, two pianos, and the timpanist are onstage; oboes, bassoons, and the organist are in the organ loft at the rear of the stage; some percussion instruments are at orchestra level, in side boxes; the entire bass section and a jazz drummer are in the first balcony; and three piccolos and three clarinets are on one side of the second balcony, with a glockenspiel and xylophone on the other side. The work was inspired by, but is not descriptive of, a day spent passing through a field of icebergs while the Brant family was on a trans-Atlantic voyage in 1926.

Karlheinz Stockhausen's *Gruppen für drei Orchester* (*Groups for Three Orchestras*) (1957) is scored for three orchestras, each comprising sixteen strings, six woodwinds, seven brass, and six percussion, and each with its own conductor. The orchestras are located to the right, left, and front of the audience. The effect achieved is the movement of sound, not only straight out from each orchestra toward the audience but also from orchestra to orchestra across the performance space. The orchestras sometimes merge in accelerations and crescendos, but they then separate and, at different tempos, "fling" sounds at each other, going in a circle. The three groups come together again for a grand, concluding brass chord that whirls around the heads of the audience (Harley 1993, p. 133). Thinking that four orchestras would be better than three, Stockhausen (see Page 1999) then composed *Carré* (1960) for four orchestras and choirs, with four conductors. The groups are located at the four middle points of the sides of a square. Each group is static, to give the listener, in Stockhausen's (1964, p. 103) words, "a little inner stillness, breath and concentration."

Iannis Xenakis (1922–) went a considerable step further. For *Terrêtektorh* (1965–1966), he has the audience in a circular space. The conductor is in the very center, the musicians are distributed according to a mathematical plan within the audience, and three percussion groups are located in different areas of the periphery. In Xenakis's words (1971, p. 237), "The orchestra is in the audience and the audience in the orchestra." The audience senses movement but possibly also a cacophony, for, as its members are immersed in "ordered or disordered sonorous masses, rolling one against the other like waves," the sound begins to swirl. It seems likely that, due to the distribution of the various instruments relative to each member of the audience, each audience member will hear different emphases; hence, this work can have no one definitive "sounding."

Performance spaces have been manipulated in other ways to try to create different sound combinations, but hopefully these several examples demonstrate adequately that, while the standard distribution of orchestra members within a performing space is closely followed by orchestras around the world, some spatial variations are tried.

PERIPHERAL SOUNDS

Some settings give us *peripheral* (and others, *encircled*) listening, inasmuch as the human or nonhuman "performer" and the listener are located outside, in "nature," surrounded by all sorts of sounds. Sounds in the open environment can come from any source. The sounds at once surround or encircle us and, yet, not being directed specifically at us, are peripheral to the degree that they varyingly intrude upon our consciousness. If barely audible, they may be ignored; if loud, they may not be missed.

We owe it to ourselves to be observant of the many peripheral sounds that encircle us each day. One trouble is that as we become accustomed to a place, we put "nonessentials" out of mind, so to speak, or, at least, we become highly selective in what we "see" and "hear." This is so because with familiarity our environment becomes commonplace. We take many things for granted and, thus, in a way, do not see or hear them. We become highly selective, walking particular routes and ignoring others, seeing some landmarks and hearing some sounds but bypassing or ignoring others (Knight 1983). As a remedy to this all-too-standard behavior, it is useful sometimes to try to be a stranger in your familiar locale, to play tourist and to walk around as if doing so for the first time. We may see and hear things we take for granted in a new light.

Some sounds outside are, in a sense, directed and focused, though the sound that I'll now identify was also, at one and the same time, peripheral as part of the encirclement of sound within which my wife and I lived. Our Chicago apartment was one block from the "L," the elevated commuter train. At first, the rumble of engines and screaming of metal wheels taking the curves were impossible to ignore, but in time we developed an insensitivity to the sounds, and they became a mere minor inconvenience. But what did we miss? Did we do the same with bird sounds? In contrast, when we are in rural Vermont, we hear all sorts of peripheral sounds that constantly give us cause for pause and reflection. On calm days, we can hear the astonishing array of songs by the hundreds of birds, the occasional snort of deer and moose, the far-off cry of wolves, and the periodic bark of the watchdog at the house across the valley. Human sounds intrude, notably the occasional car and truck, occasional, faint, joyous shouts of the boys from "the place up the road" as they explore our woods, periodic gun shots as people enjoy target practice in the quarry a mile away, barely audible engines on high-flying planes, a tractor engine in the distance as John cuts the hay, and, each evening, from about four miles away, the muted hoot of the engine hauling the train from New York just before it arrives at the Cox Brook Road crossing. Wind and rain bring different sounds, sounds that charm and sometimes alarm, such as during a storm when we may hear the crack and crash of a tree falling. The place never lacks interesting sounds.

We sometimes add music to the above-noted sounds, by using our radio or record and CD players. To some this is blasphemous, for nature's music alone is beautiful enough, but many thoughts can fill the mind when, for example, one listens to Beethoven's Sixth Symphony's storm movement during a passing summer storm. Equally, listening to Bruckner's Ninth Symphony on a calm, peaceful, moonlit night adds a special dimension that can stun the emotions and charge one's sense of spirituality. Listening to a Bach suite or Antonio Vivaldi's *Summer* while sitting in the garden on a hot afternoon amid nature's myriad sounds creates a marvelous interplay between a composer's perfectly structured sounds and nature's seemingly random sounds.

If we open ourselves to sounds around us, we can come to realize that the environment is a source of sounds that can be amazing in their simplicity or their complexity. Sounds in the environment can feed our souls, though some can agitate and annoy if we let them. But if we ignore all sounds, if we shut ourselves off in a hermetically sealed building, we can starve our souls! I am no psychiatrist, but I am confident that people need sounds, human made and otherwise. People with hearing impairments likely lose an important dimension of humanness, but surely they also gain sensitivity to other aspects of being human that may elude those with their hearing abilities intact. It is highly significant that one of the world's best solo percussionists, Evelyn Glennie, has a profound hearing impairment. She performs with orchestras by feeling the musical beat that comes to her bare feet through the wooden stage. She has fine-tuned one sense as a substitute for another. We can surely all learn from her courageous triumph over a disability. It is wonderful that Glennie, who can enjoy "visualscapes" but cannot hear soundscapes, can nevertheless create the latter for the enjoyment of those who can hear.

URBAN SOUNDSCAPES

When we stop and listen within an urban setting, it is important to try consciously to hear the total soundscape. Soundscapes are part of the total urban fabric. Noises from all sorts of sources surround city dwellers. With imagination, these can be "read" and enjoyed, not just heard or shut out. Some places have special sounds, such as subway, bus, and train stations; parks; and sports grounds. Some sounds heard involve musical instruments. Street musicians of varying abilities, on flutes, guitars, drums, trumpets, violins, or other instruments, provide music for passing people in the hope of money in return. But a street musician must compete with the scuffle of shoes, the hum of talk, the caw of blackbirds and "cull" of seagulls, the whistle of wind coursing around the corner of a building, a barking dog, the rumble of commuter trains, squealing car tires, roaring truck engines, wailing sirens, bland Musak emanating from shops, a police officer's whistle, honking horns, the "brrring" of bicycle bells, and the roar of planes flying overhead. With all that noise, street musicians may be barely audible; thus, they have to be strong willed so as to not get upset that their performances are but one sound among many. But their music, no matter how loud or low, how well or poorly played, can nevertheless effectively touch weary souls and stimulate smiles.

Some street musicians are quite noticeable due to the unexpected "noise" that is forthcoming. Such is the case in Cambridge, England, where for quite some time a one-man band has regularly performed in the City Center on Saturdays. His tootles, whistles, bangs, and crashes join together, as his legs and arms pull and push, and his mouth moves from instrument to instrument, all to produce, via odd combinations, a unique amalgam of sounds that are a distinctive part of the city's soundscape.

To live with urban noise all the time, without filtering, could drive us mad. But in filtering, do we leave ourselves open to surprise? American Max Neuhaus is a pioneer in this realm. He has been called an "environmental," "site," and "public" composer (Rockwell 1997, p. 145). His most public work likely is the project he set up in Times Square, in New York City, where the sounds he created using electronic equipment emanated from beneath a grate in the roadway. Why would he want to add sounds to one of the busiest and noisiest urban landscapes in the world? Neuhaus sought to challenge the city's noise dramatically by creating a pleasant, rich organ chord that droned endlessly on a twenty-four-hour basis. Some passersby did not hear it, at least consciously; others did hear it and were puzzled, pleased, or amused.

The idea of altering urban sounds by the intrusion of a surprisingly pleasant sound raises the question of what can be done to introduce silence to the city. Cities are generally very noisy places. Some high-rise-building designs have taken this into account, providing rooftop gardens and courtyards to which people can escape. These spaces give relief from street noises and from high winds created in the "tunnels" between buildings. Some inner-city gardens function well as quiet places due to the placement of trees on the periphery and within the grounds. The numerous public and private gardens in London demonstrate the importance of small, green urban gathering areas. Many of these gardens are a full city block in size. Apart from some seats adjacent to a path, the space is usually empty of structures. Located just a few feet from the general-public (sidewalk) area, the gardens represent havens from the hustle and bustle of daily life. The drop in sound due to strategically placed peripheral vegetation can be quite dramatic. Birds normally not heard on the street may be heard in such gardens, even though people and vehicles are passing close by. Also to be heard may be a strummed guitar or some children singing or shouting in delight as they play with a ball or skip rope. Those gardens are havens for relaxation and contemplation.

AN URBAN MUSIC GARDEN

Cellist Yo-Yo Ma felt that not only should there be such quiet spaces in all cities but that music should be an essential ingredient of their existence. He and colleagues tried to get approval and financial support for such a place in Boston, Massachusetts, but the effort failed; in contrast, the city of Toronto, Ontario, in Canada jumped at the idea. Thus, since June 10, 1999, there has existed the two-acre Toronto Music Garden on the Lake Ontario shoreline (figure 8.1), immediately southwest of the highest structures in the city's downtown. Inspired by mental images evoked by Johann Sebastian Bach's *Six Suites for Unaccompanied Cello* (ca. 1720), Ma, landscape designer Julie Moir Messervy, and local artists created a fascinating place that is an explicit interpretation in nature of Bach's first suite (figure 8.2).

Figure 8.1. Looking East along the Water's Edge Promenade, with the Toronto Music Garden on the Left.

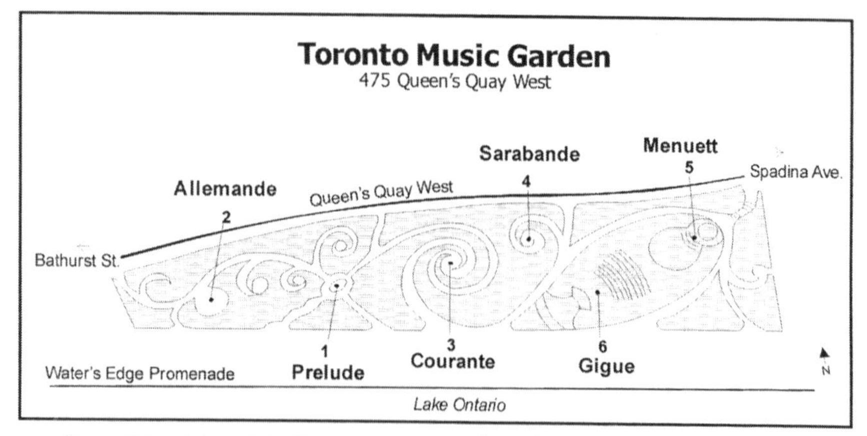

Figure 8.2. Map of the Toronto Music Garden, Showing the Six Musical Areas.

- "Prelude" captures the feeling of a flowing river (figure 8.3). Granite boulders from the Canadian Shield are placed to represent a streambed, with low-growing plants along its banks. An alley of trees with straight, regularly spaced trunks suggests measures of music.
- "Allemande," an old German dance, is expressed in the landscape by birch trees along an inward-spiraling path with seating along its

Figure 8.3. Looking toward the Core of Prelude, Toronto Music Garden.

Figure 8.4. The Allemande Entrance to the Toronto Music Garden.

route for moments of contemplation and a circle of Redwood trees at a high point through which the harbor can be seen (figure 8.4).

- "Courante," originally a dance form found in Italy and France, is expressed on the land by a footpath that forms a large, spiraling swirl through a small field filled with flowers that attract butterflies and birds, and topped by a maypole.
- "Sarabande," an old Spanish dance, is expressed by another path in the form of a small, inward-arching circle, enclosed by tall conifer trees. There is a small, stone stage for poetry readings and musical performances, as well as a small pool that reflects the sky.
- "Menuett," a seventeenth-century French dance, is revealed in the landscape by a circular pavilion, with precise symmetry and geometry, to shelter small music ensembles and dance groups.
- Finally, the landscape imprint of "Gigue," a rollicking old English dance, involves giant, grassy steps that curve around an amphitheater with a stone stage, a place for performances, and on upwards to offer views of the harbor.

Shrubs and perennials abound during summer months in many parts of this fascinating landscape. When the garden was dedicated, solo cellists (one being Ma) performed Bach's suites. On any day, people escaping from office settings are able to enjoy informal performances. The

Toronto Music Garden is public space, with free admission year round. Designed and built expressly to reveal music in landscape, this garden is a remarkable urban feature and an ever-changing soundscape. As such, it is a place for therapy.

MUSIC FOR THERAPY

The Toronto Music Garden can be therapeutic for someone troubled by stress or overwhelmed by sounds of the surrounding city. The combination of the place and the music being performed on an ad hoc basis can be meaningful for many visitors. As noted in chapter 1, R. Murray Schafer coined the term *soundscapes* to refer to our sonic environments, within which there is an ever-present array of noises, some pleasant and some not, some loud and others quiet, which we choose either to hear or ignore. He suggests that we need to take "soundwalks" to "ear clean" ourselves, for we take in so much, even when we try to ignore the sounds in our environment, that our minds are filled with noise clutter. I like to use a computer term, *DEFRAG* (from "defragmentation" of files on a computer disk), with respect to our normal need to untangle ourselves not only from our daily and weekly overload but also from the sounds around us. Yo-Yo Ma's music garden in Toronto is just such a place where one can go to DEFRAG while enjoying some quietness and whatever music an artist happens to be playing. To appreciate fully whatever the sounds are around us—or simply to enjoy silence, if that is what is "there"—we need to be open in order to listen and truly hear.

Music is used in other situations to help heal people disabled in some way by a stressor, such as a physical or mental ailment. *On Deadly Infections*, attributed to Democritus, states, "Snake bites are cured by music of the flute when played skillfully and melodiously" (Meinicke 1948, p. 81). To my mind, it would take an *extremely* skillful player to take my mind off a snake bite! But ancient writings from East and West include numerous mentions of the use of music in relation to healing, so perhaps my modern, Western skepticism prevents me from accepting music's full healing power, although, as noted in chapter 7, I have experienced the power of music as a healing agent, when I sought to deal with the shock of a suicide. The aged organist and missionary Albert Schweitzer, after playing a Bach toccata and fugue, said he "felt restored, regenerated, enhanced. . . . Music was his medicine" (Cousins 1979, p. 82). And Englishman Lionel Joyce (2001, p. 6), suffering from severe depression and considering suicide, upon the advice of his psychiatrist listened to *Kindertotenlieder*, Gustav Mahler's songs on the death of children, and to Richard Strauss's *Four Last Songs*. "The sound of this music confirmed the truth about the joyless and desperate bleakness of life. But someone else knew what I knew and spoke in my terms." He then continued to listen to music, and with professional help, his condition gradually improved. For many, the importance of silence is supreme.

ON SILENCE

There are times when we need to remind ourselves of the importance of silence, for silence "is the essence of sound" (Schafer 1977, p. 259). Indeed, music depends on silence: silence within sections of a movement of a work, be it a "G. P." (grand pause) of perhaps one bar's length or a rest between notes. A composer generally places silence between movements when he or she hopes we will consider what has just been played as a prelude to what is to come. If we listen to silence, we may hear things we have not heard before or things that we had forgotten about. John Cage has made this point strongly. His *4′33″* (1952) is a delight. Imagine the following: twenty-five musicians—or two musicians—walk onstage and sit down. The leader signals the start, but no one onstage moves. The performers do nothing. They just sit. Some members of the audience quickly get fidgety, but others welcome the silence, curious though they may be. Frowns and smiles are evident among the audience. As time passes, everyone present begins to hear sounds, from those present (a program dropping on the floor, a cough, or a cough drop being unwrapped), from the hall itself (a creak here, a banging pipe there), or faintly from outside (a passing bus, a distant jackhammer). Then, at exactly four minutes and thirty-three seconds, the leader signals and the performers stand and take a bow, then walk offstage. To explain, the work is marked "*tacit* for any instrument or instruments." "Tacit" means silence and, in a musician's score, "do not play." Cage's point is that there are always environmental sounds worth hearing and contemplating. By paying attention to ambient sound, we can better train ourselves to hear, an act that is a level higher than just listening (Schafer 1977, 1986). Of course, another "hearing" of Cage's work will provide a quite different experience since the surprise and suspense will be absent and the ambient sounds may be quite different. The changes will in no way undercut the experience, however; rather, the experience may be richer.

ON A LAKE SHORE

R. Murray Schafer's music often incorporates sounds in nature because many of his compositions are performed outside. For example, his *Music for Wilderness Lake* (Schafer 1984, pp. 95–100), was written for and performed by twelve trombonists located at various spots around a remote Ontario lake, in a location without vehicular- or air-traffic noises. Each performer was invited to interact with and reflect upon his or her assigned location prior to playing. From wherever a player played or someone listened, the time lag for each instrument was slightly different, as, of course, was the direction of the sound. Schafer determined that since sound travels at just over 330 meters per second, it would take three seconds for players on one shore to hear the sounds from the opposite shore and proportionally less with decreasing distance around the shore to wherever the listener was standing. A small, floating platform was placed in the center of the lake, and from there the work was conducted and recorded. Dawn and dusk in late spring were determined to be the best times for performances since those are the peak times at which birds sing, and birds were important participants. The performance was wonderfully alive, combining the natural and human-made musical landscapes. Interestingly, the work has also been performed on the Amstel river in Amsterdam, where the surrounding urban sounds were utterly different from those at remote Wilderness Lake.

THE PRINCESS AND THE STARS

Imagine driving for several hours to reach a forest reserve, where you set up your tent at the base camp near the park entrance so you can stay overnight. You share a common meal with thirty or so other folk. A short talk takes place about the next morning's transport plan, and then most people wander off to their tents. At 3:00 AM, some people can be heard getting up. Later, we learn that they are the performers. At 4:00 AM, the remainder of us are awakened and, warmed with coffee, loaded into two small buses that bump along a logging road and, so, take us deep into the woods. Little can be seen, other than the snatches of trees fleetingly illuminated by the bobbing headlights. The buses stop, and with a few flash lights guiding the way, we are led down a twisting path to a small cluster of wooden benches located beside water. No lights can be seen apart from the twinkling stars. The moon must have come and gone. The night air is cold, so it is just as well that we were warned to wear heavy winter jackets. Our requested silence is broken only by lapping water and the occasional cough or groan as the roughness of the benches begin to take their toll. We sit for about half an hour, absorbing the sense of the dark place and wondering what will appear before us when the sun comes up. Our eyes adjust, and we can just make out the outline of a lake. We continue to wait in silence, absorbing the feel of the dark place.

A solo soprano's voice, that of the princess, who is never seen, drifts across the water. Her unaccompanied aria gives us shivers, so mysterious is the moment. Other voices join in, as do various instruments, including a cello, trumpet, trombone, and some percussion. The sounds seem to come from many directions, but it is not possible to tell accurately from where. Sometimes they seem to come from somewhere on the shoreline, at other times, from high up in the wooded hills that rise from the shore. From several directions, four huge, lantern-lit canoes are rowed slowly into our line of vision, singers and dancers on board, relating a story in song and in movement. In each canoe is one of the four key characters: Narrator, Wolf, Three Horned Enemy, and Sun. The singers sing, and the musicians continue to play from around the shore. The different locations, the twists and turns of bays, the varying elevations, and the moving canoes lead us to hear echoes and overlaps that form a fascinating interplay as we listen. Then, as if by magic, the first flash of sunlight appears, just as the birds begin to sing. The birds—called Dawn Birds in the script, we learn later—are newly awake and readying themselves for their day of activity and, in doing so, have become part of the performance. The long, forlorn call of loons (a water bird) can be heard coming from different directions. Other birds join in. We wonder if a wolf will howl (as happens on some mornings) or, perhaps, a bear will grunt (but the thought is quickly put out of mind since, if we heard a bear grunt, that would mean it was already too close). The boats continue to move around, singers singing, dancers dancing, and players playing, until all withdraw from sight and sound making, and we, the audience, are left to reflect on what we just experienced.

The Princess of the Stars is the prologue to the *Patria* series (Schafer 1984, pp. 101–105; 2002). The other works in the series similarly demand an environmental setting. Performances take place on the shore of wilderness Wildcat Lake, located in the Haliburton region of Ontario (it has also been performed at Two Jack Lake in the Canadian Rockies). The experience is a very physical one. It could not be repeated with any degree of similar success if performed in a concert hall or, indeed, within any structure. For example, the echo of the voice of the princess was sometimes louder than the initial sound, and sometimes the echo was repeated several times before it reached us. The light changes caused by the rising sun were wondrous, as we gradually saw the emergence of lakeside details, announced as they were by both performers and birds. We had experienced music *in* landscape, with the sounds of nature forming an integral part of the experience. The effect of stationing the singers and musicians in various locations around us, close by and far off down the lake, led us to hear sounds at different times, overlapping and fading in fascinating ways. Clearly, the experience of the performance in that place was integral to the outcome of the performance.

MUSIC IN STADIA

Anyone who knows the fine game of rugby appreciates that whenever any team plays against the Welsh national side in the Welsh national stadium in Cardiff, the home team begins with a vast emotional advantage due to the crowd's singing. The sound is amazing, with sixty thousand and more voices raised in song—in harmony! It gives opponents a bad case of the jitters. Of course, when Wales plays England in Twickenham, in London, they have to contend with the thousands of English fans singing the borrowed Negro spiritual, "Swing Low, Sweet Chariot." Some prerugby game rituals are war based. In Murrayfield in Edinburgh, visiting teams facing the Scots must deal with the playing of the bagpipes. And whenever the New Zealand "All Blacks" (the national team) play, opponents must face the terror of a Haka, a fearsome, chanted, Maori war dance. And at all international games, the national anthems of the two teams playing are performed, usually by live military bands, and sung by the assembled crowds, which can both lift or sink the emotions of teams who must stand on the ground until this ritual act is completed prior to the kickoff.

The atmosphere of international rugby games, and indeed, of club and regional games too, can be fascinating to experience. A composer who fell in love with rugby and the spirit of the game and of the crowds was the Frenchman Arthur Honegger. He wrote that he preferred rugby to football (soccer) because "it is more spontaneous, more direct and closer to nature than football which is a more scientific game . . . for me the savage, brusque, untidy and desperate rhythm of rugby is more attractive" (quoted in Spratt 1987, pp. 148–49). Countless numbers of Frenchmen, New Zealanders, South Africans, Scots, and others around the globe would agree with him. But that apart, Honegger's *Rugby* (Mouvement symphonique no. 2) (1928) is a fascinating expression of the spirit of the game. He wrote, "All it does is to try to express in my musician's language, the attacks and ripostes of the game; and the rhythm and color of a match at the Colombes Stadium" in Paris (quoted in Spratt 1987, p. 149). Listening carefully, one can hear the two scrums pushing hard, as one team tries to steal the ball from the other team. At another moment, one hears what could be one of the back lines happily watching one of the players running, as if in slow motion, toward the goal line for a try. The representation of the typically good-natured rugby crowd is a joy to hear. It is not known if this work has ever been performed outside, in a rugby stadium, but its inspiration can be heard every weekend during winter seasons in stadia around the globe.

During the half-time break in some American football games, huge bands of brass, woodwind, and drum performers play and march in patterns around the field. The ritual is an intriguing one. The bands sit in the stands during the game. Each band encourages both players and their supporters by playing "their" team's song, which, of course, must be "answered" by the opponent team's band and its supporters. These spectacles of music in places can be quite impressive and are sometimes more interesting than the games played on the field by the helmeted and well-padded players.

Some sports stadia have been used for particular performances of music. The massive crowds in London's Wembley Stadium for rock concerts are legendary. The nearest equivalent in the orchestral world may well be the huge crowds attracted to hear the two concerts associated with World Cup football (i.e., soccer) finals. Luciano Pavorotti, Placido Domingo, and Jose Carreras or, as they are popularly called, the "Three Tenors," attracted worldwide attention during the 1990 World Cup when six thousand people filled the Roman Baths of Caracalla (built between AD 212 and 216) and more than eight hundred million others watched on television. During the 1994 World Cup, 56,000 people were in Dodger Stadium in Los Angeles, and 1.3 billion people watched on television as the three tenors sang, again conducted by Zubin Mehta. Ignored in all of the hoopla were the orchestras: the joint orchestras of the Rome Opera and the Florence May Festival in 1990 and the Los Angeles Philharmonic in 1994. If noting record numbers is important, then the largest single outdoor concert may have been that given in New Zealand by Dame Kiri Te Kanawa in the Auckland City Reserve, a vast, open parkland south of the city center, when half a million people were present for a "welcome home" (from Europe) performance some years ago. Of course, that was small potatoes compared to "her" huge radio and TV audience when she sang at the wedding of Prince Charles and Lady Diana Spencer (after the ceremony, Diana, Princess of Wales) in London's Westminster Abbey.

OUTDOOR FESTIVALS

The idea of outdoor performances is ancient. Consider, for example, the performance of plays and other events in outdoor amphitheaters in ancient Greece. Centuries later, in 1717, barges conveyed more than fifty musicians on the Thames as they played George Frideric Handel's *Royal Water Music* for King George I as he was rowed from Whitehall to Chelsea and back. And then, in 1749, the *Music for the Royal Fireworks* was performed before the king and an assembled crowd of twelve thousand. That first performance included more than one hundred violins, twenty-four oboes, twelve bassoons, nine trumpets, nine horns, one contrabassoon, a serpent (a bass cornet about eight feet in length that is folded back on itself), and three pairs of timpani. Unfortunately, shortly before the conclusion of the fireworks show, some of the pyrotechnics became uncontrollable, and the huge (one hundred feet high and four hundred feet long) launching structure burned, leaving two people dead and many others injured. Thankfully, the Royal Library, just five hundred feet away, did not catch fire!

Outdoor performances remain popular. Notably, since the mid- to late-nineteenth century, band shells in numerous parks in Britain and the countries of the Commonwealth, as well as in U.S. town squares, are used by pipe bands, brass bands, concert bands, and sometimes orchestras to perform on summer Sunday afternoons and evenings, with the audience sitting on folding chairs and blankets. Such events take music to people who might not otherwise be able to hear a concert. Some external sites are grand, with hundreds or even thousands of people seated on facing hillsides, listening to performers of all stripes, including full symphony orchestras.

Many major orchestras perform in outdoor locations each summer. In the United States, these include Tanglewood (Boston Symphony Orchestra, Massachusetts), Ravinia (Chicago Symphony Orchestra, Illinois), Mann Music Center, formerly known as Robin Hood Dell (Philadelphia Orchestra, Pennsylvania), Blossom Music Center (Cleveland Orchestra, Ohio), Wolf Trap (National Symphony Orchestra, suburban Washington, D.C.), Meadow Brook (Detroit Symphony Orchestra, Michigan), and the Esplanade (often shown on public television in the United States) that sits beside the Charles River (Boston Pops Orchestra, Massachusetts). Many others exist, including some for orchestras created annually for summer performances (such as the Grant Park Orchestra, which performs in a park adjacent to Lake Michigan near downtown Chicago).

FESTIVALS: LARGE AND SMALL

Tanglewood Music Center is a particularly famous summer orchestral location. Its incomparable landscape is surely the standard against which all other permanent, orchestral, summer-performance settings are measured. The landscape is beautiful, with a magnificent mix of lawns and trees and an unparalleled view of the countryside and a lake. There are some residences and buildings for rehearsals and orchestral and chamber music performances, opera, jazz, and theater. Other buildings are for administrative needs, cafeterias, and shops, and there are several huge tents in which visitors can socialize while enjoying drinks and food.

Located just outside the town of Lenox in western Massachusetts, and more or less due west from Boston and north from New York City, Tanglewood is one of the world's most influential music festivals. It was founded in 1934 as the Berkshire Symphonic Festival, with the New York Philharmonic performing during the first two seasons. Thereafter, the festival committee invited Serge Koussevitsky and the Boston Symphony Orchestra (BSO) to participate in the following year's concerts. After a successful run of concerts, the BSO was gifted the Tappan family estate of 210 acres of lawns and meadows. The site was named the Tanglewood Music Center and the festival was renamed Tanglewood. During a 1937 concert, there was a major storm; rain and thunder twice interrupted the performance of Richard Wagner's *Reinzi Overture* and forced the cancellation of his whisperlike "Forest Murmurs" from *Siegfried*. One of the founders of the festival stood up at the intermission and persuasively suggested that a permanent building was needed to shelter the orchestra and the audience. Support was immediate, and soon thereafter, the Shed was constructed (Kupferberg 1976, pp. 55–60). This huge structure, open on three sides, holds five thousand audience members. In 1986, the BSO acquired a neighboring property of 120 acres. With other lands added, Tanglewood Music Center now totals 575 acres. What makes this landscape so special?

Tanglewood serves as the superb summer location not just for the BSO but also for an annual summer program to which about 150 young musicians from around the globe (who are already skilled instrumentalists, vocalists, conductors, and composers) are invited for intensive study with BSO musicians and specially invited artists. It is surely a good measure of Tanglewood's importance that 20 percent of the members of U.S. symphony orchestras and 30 percent of all principal players in the United States have studied there.

The Tanglewood Music Center Orchestra (TMCO) includes about 100 (out of the 150) young musicians who annually perform under the baton of the BSO's music director and other conductors. Within just a few days of first convening, the TMCO reaches a very high standard; so fine are the individual musicians that they respond magically to the conductor's directions. Beyond their level, however, there is the BSO, and it too gives regular performances. During the course of a summer season, more than three hundred thousand people attend the concerts. During a three-day period at which I was present, more than forty-four thousand attended three concerts. Where did they all sit? Five thousand were in the Shed: high-priced-ticket holders got to sit near to the stage; the cheaper seats were behind them. Many others sat on folding chairs or blankets on the huge lawn, often enjoying their food and wine as they awaited a performance (figure 8.5). On Friday evenings, prior to the main concert, a free chamber concert is held in the magnificent nearby Seiji Osawa Hall, also designed for an audience seated both inside (nine hundred seats) and outside (many hundreds

Figure 8.5. The Shed at Tanglewood.

of people sit on the large adjacent lawn). During and after the chamber concert, the grounds of Tanglewood are covered with people enjoying picnic suppers. During the main concert, two giant screens show the outside audience what is happening onstage, as they listen with ease to the performance via speakers (figure 8.6). When there are no concert performances, one can wander the grounds and listen to the sounds of singers and solo or chamber groups of musicians emanating from the several small buildings, as well as, sometimes, the BSO or TMCO rehearsing in the Shed. Tanglewood truly is a landscape of music!

Not all summer locations are grand in size. By way of an example, the Elora Festival in the village of Elora, Ontario (an hour's drive west of Toronto), with just 3,500 inhabitants, has for twenty-six years held its six-week summer festival of singing, orchestral, chamber, and solo performances in two of the village's churches, a flooded quarry, and the highway department's large wooden gambrel barn (used for salt storage during the winter). The barn has remarkably good acoustical qualities for both solo artists and large orchestral and choral performances of, for example, Henry Purcell's *The Fairy Queen*, Felix Mendelssohn's *Elijah*, and Giuseppe Verdi's *Requiem* (figure 8.7). A nearby, flooded quarry is sometimes used for vernacular, generally folk performers. Some performances have taken place on an anchored platform in the middle of the water with the audience around the quarry's rim (figure 8.8). Over ten thousand people

Figure 8.6. One of Two Video Screens and Some Speakers on The Shed at Tanglewood.

Figure 8.7. Inside the Huge Wooden Gambrel Barn in Elora, during a Performance by the Elora Festival Choir and Orchestra. (Photo by Vern McGrath, used with permission from the Elora Festival.)

Figure 8.8. A Group Performing on the Floating Stage in the Flooded Quarry, Elora Festival. (Photo by Vern McGrath, used with permission from the Elora Festival.)

annually attend the festival, attracted by nationally and internationally renowned artists, the superb professional Elora Festival Singers (recorded by Naxos International), and the festival's symphony orchestra. The audiences are from the surrounding local region, the Toronto, Hamilton, and Kitchener-Waterloo metropolitan areas, and farther afield, including the U.S. states of New York, Ohio, and Michigan. In contrast to Tanglewood's huge size, the Elora experience is intimate.

Classical music festivals are to be found on all continents, though throughout Europe and North America especially they are "ubiquitous events" (Waterman 1998a). It is important to note the use of the word "festival," for it implies celebration and a positive experience. Some festivals are actually better described as festivals within festivals, for they include different types of musical experiences. The Elora Festival, for example, uses its Quarry Concert Series to attract people not interested (or also interested) in classical music, while the remainder—the core—of the festival takes place indoors, in the churches and in the remarkable barn. The local Anglican vicar offers free tea to concertgoers between concerts on Sunday afternoons! While some local residents choose not to attend any performances at all or, perhaps, to go only to the free "popular" concert for locals or the effervescent "The Last Night at the Proms" performed in the gambrel barn, they nevertheless take pride in knowing "their" village has the festival, and this pride forms part of the villagers' sense of place (Knight 1998).

The Lenox and Elora business communities are happy to have the economic benefits associated with the festivals. Tanglewood is to a degree an elite experience, judging from the obvious wealth of some of the patrons (based on cars in the immense parking lots, the price of tickets nearest to the stage, and the exhibition of wealth by some of the people sitting on the lawn, enjoying obviously expensive and ostentatiously served food and wine—prior to going to their seats in the Shed). But it is clear that many in the audience either are not wealthy or do not flaunt it (judging from other cars in the parking lots and the number of simple sandwiches). From an economic perspective, Tanglewood is one of many cultural attractions in the Berkshires in western Massachusetts, a region that attracts numerous New Yorkers. The festival is an excellent economic generator, as indicated by the exorbitant rates charged by hotels and motels during the orchestral festival.

CARNIVALS FOR ELITES?

Stanley Waterman (1998a) wonders if music festivals are simply "carnivals for elites." Some festivals, such as the small Elora Festival and, across the Atlantic, the huge Edinburgh Festival, seek to bridge "cultures" between the "popular" and "high" arts, whereas others clearly target one audience or the other. Waterman makes the point that a music festival is a "fleeting experience." I would argue that this conclusion applies to all performances, whether by a rock group or a major symphony orchestra, though, of course, the experience of having heard a particular piece of music or seen a performance by a particular group can have a lasting impact on an individual. Some music festivals have little landscape impact (as in Elora where existing public and religious facilities are used) since, once the festival is over, the place returns to its normal functioning, with the only reminder being old posters or perhaps posters for the next year's event. Other festivals have a long-lasting, major impact on the landscape, for they encompass vast landscapes that have special buildings—the orchestra's permanent sound shell—and a setting for the audience. Some may include facilities for teaching, rehearsals, and performances; perhaps some residences; and manicured lawns. Such places include Tanglewood, already discussed, and Bayreuth in Germany, where Wagner's estate is located, and also the Festspielhaus, a magnificent opera house where the annual Bayreuth Festival is held. The latter is regarded by many as *the* place to see performances of Wagner's huge musical canvasses, even though, as Rolf Sternberg (1998) has shown, his works quickly diffused from Bayreuth to many other locations (see also Millington and Spencer 1992). But we should not lose sight of Waterman's concern. Is it the case that only an elite audience (which is well educated in classical music matters or maintains a certain social standing by being seen to attend) is willing to pay the high price for entry to such festivals as Salzburg or Bayreuth and to sit through many hours of orchestral and other performances, let alone a Wagner opera? Such certainly seems to be the case, though, of course, there are also the large numbers of people who are willing to attend reasonably priced festivals, such as in Elora, and the large-crowd, outdoor, musical functions offered each summer by the Berlin Philharmonic, the Philadelphia and Chicago orchestras, and many like organizations.

ELITES AND CULTURE

In our current age of sound overload, the "problem" of decreasing attendances (for some, but not all orchestras) may not be the fault of orchestras and the music they perform but of their association with a hall where, it is believed, only the well-dressed, educated "elite" go. Today, orchestras are developing innovative approaches to people in order to attract new audiences. As an example, we can note that since the issue of dress was identified, "dress-down" concerts have been advertised; thus, the landscape impression of dressed-up audiences is absent, or decreased. It is now common to find people at regular concerts who are casually dressed and feel quite comfortable about it. Some elites still dress up, perhaps partially mirroring the British elites who attend the Grand National horse race for relaxation among like-minded people—and to be seen. This point brings us back to Waterman's observations.

Waterman maintains that the continuing importance of festivals is essential for sustaining culture. Drawing upon his observations of the Kfar

Blum festival in Israel, Waterman (1998b, pp. 253, 261) writes that, initially, it was "dominated by audiences of elite groups who gave it a specific character and made it a highly desirable social event." The festival became, in a sense, the focus of an annual pilgrimage, for people enjoyed staying for a relaxing holiday, while also listening to music. In time, the festival became "a part of the social and cultural calendar," for the elites, by their actions, "unintentionally transformed the festival into a commodity," to be "used" once a year (Waterman 1998b, p. 261). Rather than remaining as "just" a pleasant place for the celebration of music, both the festival and its "place" (a once quiet kibbutz in northern Israel) were transformed (the kibbutz into an active place for the arts, with, for instance, a 1967 guest house's transformation into a thriving hotel). A core issue is what music should be celebrated, "cultivated classical music [that] is also a music of exclusion," to quote Waterman (1998b, p. 263), or other forms of music that would reflect the complexities of music in relation to various ethnic and cultural identities? Mention of this case is useful for it reminds us that some classical music can be exclusionary. It also is useful for what it suggests about the issue of origins and dispersals and cultural conflict due to convergence.

FESTIVALS IN MUSIC

The spirit of festivals has been captured in music. I have already discussed Ottorino Respighi's *Roman Festivals* (chapter 4). Another well-known work focuses on a parade. Claude Debussy's *Fêtes* (*Festivals*) (1899), after a brief, rich, pulsating orchestral opening, suddenly stops, and as if from far off, we hear the beginning of a distorted march. The music gets increasingly louder to suggest that a group of players is marching toward us. A climax is reached right when the group arrives, and then a gradual diminuendo occurs, as if the band were marching away from us into the distance. The work ends with a barely audible cymbal crash. During its performance, the array of colors is constantly altering. The music changes tempo many times; at the close, the tempo is almost funereal. Debussy was more interested in creating a mood or atmosphere associated with his topic; thus, *Fêtes* is more about color creation and the mood of the event than it is descriptive. Using the "carnival" word link suggested by Waterman, we can include in this discussion Robert Schumann's miniature "character piece" (i.e., the portrayal of a single idea, mood, or emotion) called *Carnival* (1833), which describes a Mardi Gras ball. The latter, of course, is a festival of a sort. The music, for piano, is arranged as a set of twenty short pieces, each representing a character at the ball. The characters include the clown Pierrot and the pantomime figure Harlequin, but masquerading as them are such persons as Chopin and Paganini. This musical ballroom scene is whimsical and humorous and, yet, touched with melancholy. Schumann's *Festival Overture* (1853) for orchestra and choir had a different purpose: it was written for a special celebratory occasion in honor of Rhine wine, the growing of which, of course, creates a distinctive imprint on the cultural landscape (de Blij 1983, 1985).

Let me take a liberty and digress so that I can mention the traditional musical celebration that occurs to this day in New Orleans when some African Americans die. As the funeral party wends its way toward the cemetery, a jazz band plays a wailing, slow march, which is really more of an amble. In contrast, following the burial, the band bursts into a bright song, and the funeral party quickly steps back to the home, singing, among other tunes, "Oh, When the Saints Go Marching In." This two-part expression of pain and joy reveals important aspects of belief in a public manner and creates a fascinating landscape in music. Now, let us return to festivals.

GRADUATION FESTIVALS

Festivals of a quite different type from those identified above revolve around "capping," or graduation, ceremonies at universities the world over. After their several years of study, students in many countries are known for lively celebrations just prior to their departure from university. City parades occur at some institutions, while, in other places, celebratory drinking and music making are evident, sometimes in shows. In 1880, in return for receiving an honorary doctorate from Breslau University, Johannes Brahms composed his *Academic Festival Overture*, which includes a number of German students' drinking songs and concludes with a rousing rendition of *Gaudeamus igitur*, the "identity" song sung by students in many countries. Antonín Dvořák also composed a piece to celebrate the receipt of an honorary doctorate from Cambridge University in 1891, and, following Brahms' example, he dedicated the *Carnival* overture to the university. Interestingly, his original title for the work was "Life." The work supposedly symbolizes both the active, robust and the contemplative sides of a person. However, inserted in the middle of the work is a section Dvořák called "Nature," a small motif that he also included in two other overtures he wrote in the same year. When Franz Joseph Haydn received an honorary doctorate from Oxford University in 1791, he conducted a performance of Symphony no. 92. Instead of being a celebratory work composed in recognition of his degree, however, the symphony had been written and dedicated to Comte d'Ogny two years earlier. Still, we know that symphony as the *Oxford*.

At a graduation ceremony at Carleton University, Ottawa, a number of years ago, an honorary doctorate was given to jazz pianist Oscar Peterson. Rather than offer a speech in response, he improvised on the piano. That unexpected soundscape is still treasured in the minds of the two thousand graduates, parents, and faculty who were present. At another such ceremony, the newly minted Dr. Maureen Forester sang. With one or two exceptions, graduation speeches by honorary-degree recipients have seemed mundane to me ever since!

TOURING ONE'S TURF

Some orchestral summer festivals move around, touring their general state-bound "turf" to play concerts outdoors, weather permitting. For example, the summer perambulations of the Vermont Symphony Orchestra (VSO) and, separately, the Vermont Mozart Festival Orchestra (VMFO) are planned to put them in contact with audiences normally not able to get to the regular season's performances (in the case of the VSO, centered in Burlington, the largest city, located in the northwest corner of the state). Summer orchestral programs occur in a variety of unusual locations, not least the Vermont Teddy Bear Factory, a large porch that extends from a museum, a state park, a boat club, a vineyard and winery, and a spectacular mountain-top-high site—the Trapp Family Lodge, Concert Meadow—on the grounds of a well-known inn near Stowe that is owned by the famous (*The Sound of Music*) Trapp Family (Trapp, 1990). The VMFO concludes its season at the Trapp meadow with a glorious display of fireworks to go with Handel's *Music for the Royal Fireworks*. Handel would be pleased! The locations as much as the availability of the music attract large audiences who sit on blankets and lawn chairs—and who eat their picnic suppers on the grounds beforehand. In addition, concert and brass bands play in a number of Vermont's town commons (somewhat akin to a European square in the center of a town), in gazebos in parks, at county fairs, and at town picnics or other special community events. Music is "everywhere" in the state of Vermont, it seems, as musicians take their music to their summer audiences rather than waiting for the audiences to come to them. Of course, the hope is that people who enjoy the summer concerts will then want to attend winter concert series. Smart thinking!

The New York Philharmonic (NYP) operates at a different (i.e., citywide) scale when it gives its annual, highly successful series of free summer concerts in parks in different parts of the urban area. These concerts (in Vermont and New York) are—and here I use the word "extension" loosely—extensions of 1800s and 1900s revival meetings (where gospel songs were sung), inasmuch as the orchestras remove themselves from the confines of their normal performance spaces and go to public spaces to "pitch their tents" and perform for the local population, most of whom would otherwise never hear an orchestra. American composer Charles Ives included just such a revival meeting in his *Second Orchestral Suite* (1909), complete with false starts, interruptions, melodic and rhythmic fragments, quotes from gospel songs, dissonances, and, at times, grandeur!

AUDIENCE DEVELOPMENT: RADIO AND TELEVISION

From 1927 until the invention and spread of television, radio broadcasts of classical music orchestral performances by the CBS Symphony (1927–1950), Mutual Network Symphony (1933–1943), and NBC Symphony (1937–1954) were popular (Kolodin et al. 2001, p. 829). Performances by the Philadelphia Orchestra, BSO, and NYP were also broadcast regularly. In a sense, the broadcasts were a form of festival inasmuch as millions of people across the United States listened to the several weekly broadcasts; hence, right across the country, people were learning and enjoying all at the same moment. It is instructive that people who listened to the broadcasts cut across divides based on race, age, gender, ethnic origin, and socioeconomic standing. The NBC Orchestra under Arturo Toscanini, perhaps the most famous of these radio orchestras, had a particularly large following. Due to the huge audiences, it is fair to conclude that classical music was accepted at that time as part of popular American culture. Since the 1950s, as TV broadcasts became a dominant cultural force, there has been the constant dumbing down of "culture." In the United States today, public radio, FM stations associated with universities, and some commercial classical-music radio stations continue to support orchestral music, sometimes with full broadcasts of symphony concerts, such as by the Chicago and San Francisco symphony orchestras, but these stations may be viewed as being for the educated classes, not the working classes, and perhaps not for the children of either. And there are large "holes" in the United States where no such broadcasts are available. In Canada, as in New Zealand, Australia, and Britain, government-supported radio broadcasting (including the Canadian Broadcasting Corporation in Canada [CBC] and British Broadcasting Corporation) is important for the performance of recorded, classical, orchestral music. So, at first, in the 1930s, classical music was for everyone, but with the advent of television and the demise of live radio broadcasts (coupled with the remarkable increase of other types of music stations), there has been a turning away from orchestras and classical music generally, and this presumably has meant that popular culture has lost an important ingredient.

Most radio symphony orchestras are today located in Europe, notably in Germany (Knight, in press). Only one such orchestra remains in North America, the CBC orchestra in Vancouver, British Columbia. Periodically, public television in the United States broadcasts the NYP "live from Lincoln Center," or some other orchestra, and attracts large audiences. These audiences, coupled with those assembled for free performances by, for example, the Boston Pops Orchestra at the Esplanade, suggest that great music simply is not only for a limited audience, as long as it is accessible. The broadcasts, over public television, are received also by many Canadians and people in selected other countries by satellite or cable. Audiences for these broadcasts and for outdoor performances are generally large and diverse. The same certainly holds true for the long-running U.S. radio program, *Saturday Afternoon at the Opera*, also available in Canada and many other countries. At least through their availability, which ignores the issue of ability to listen in an educated manner and, thus, best understand (as opposed to simply listening and enjoying the music for its own sake), these programs are not just for elites.

VISUAL AIDS

What do you do if you cannot afford to go to the Antarctic to see the environment and the wildlife there? As implied in chapter 6, do you rely on the symphonies by Ralph Vaughan Williams and Peter Maxwell Davies or the photographs of Herbert Ponting to provide you with images that may satisfy your desire to experience something of the place? Perhaps, instead, you can attend a symphony concert to hear a composition being played in accompaniment with a film or slides and, thus, see the environment at the same time as hearing pertinent music? Such is indeed possible, courtesy of Natural History New Zealand, an acclaimed production company, which produced a wonderful movie of Antarctic landscapes that are revealed as one listens to *Sinfonia Antartica* by Vaughan Williams. When it premiered, a giant screen hanging above the orchestra revealed the visuals, while the orchestra performed the score (now available on DVD). Natural History New Zealand's collaborations with composer Anthony Ritchie has led to other DVDs, including *From the Southern Marches* (1979) for orchestra and choir and, more recently, *Southern Journeys*, with Ritchie's music scored first (2000) and the film of some South Island landscapes created second (2001). Premiered by the world's southernmost professional orchestra, New Zealand's Southern Sinfonia in Dunedin, these works stretch the minds of armchair geographers and music lovers alike. Imaginative as they are, they are nevertheless "traditional" in the sense that they were first performed in standard orchestral halls (with a screen above the orchestra), and *Southern Journeys* can stand alone as a concert piece.

The Vancouver Symphony Orchestra has recently added visual content to its standard orchestral concerts in an attempt to attract audiences. Screens located at the front, on each side of the stage, show video close-ups of the conductor and some of the players as the performance proceeds so that audience members can see at "close range" what they otherwise would only be able to see from a distance during a concert. This method of bringing a sense of closeness between performers and audience has been taken from rock and other concerts at which large numbers of people are shown close-up enlargements of the performers. Only time will tell if this means of achieving greater intimacy during an orchestral concert will have a positive impact on audience size and whether the experiment will be duplicated in other centers. Several orchestras have tried introducing hand-held computer screens that reveal an analysis of the music being heard, but audience reaction has been mixed.

CONCLUSION

Most music is performed in buildings, sometimes with innovative spatial arrangements for the orchestra, but this chapter has demonstrated that music performance and enjoyment need not be conducted inside, for some of it can be or, indeed, in a few cases, must be performed and enjoyed outside. Some composers use sounds made "in nature" as part of the performance. Schafer, for example, chose to go outside for performances of *The Princess of the Stars*, whereas, in contrast, Olivier Messiaen (chapter 5) chose to incorporate the bird songs he wrote down in musical notations on his pad into his music, which is performed inside. Festivals are important for reaching audiences that might not otherwise hear orchestral music, though festivals also cater to specialized audiences. One hope is that people attracted to summer locations by major orchestras will consider attending the main subscription series, normally held over the winter in most climates. The point of this chapter is that music can be performed—and appreciated—in any place that an orchestra can gather, likely with an audience, and perform. Or, one can use the radio or CD player to bring music into an otherwise musicfree environment. Some places have been created for music performance and for meditation, whether or not music is actually being performed, but additional enrichment can be experienced when such places, like Toronto's Music Garden, are also places for extemporaneous performances. Clearly, music may be for all time, and music can be for all places.

9

Conclusion

Many works in orchestral music have been discussed in this volume. Except in chapter 1 and selected places in the subsequent text, it was not my intention to rehash others' ideas about music; rather, I wished to draw attention to some very special, geographic issues. In this concluding chapter, I will neither recount all that has been written in this book nor comment again on the composers, but I will highlight some points that emerged from this writing.

While some landscapes in music are fairly straightforward in their representation of some geographical phenomena, others are indirect and symbolic. Accordingly, some of this music may be heard and readily understood without any story line, while other works require an explanation if the programmatic and descriptive qualities are to be understood and appreciated fully. In contrast, musical geographies are generally quite direct, representing specific subject matter.

The process of inspiration remains elusive. As demonstrated, some composers have responded to specific geographical features (such as particular streams or lakes or mountains, as well as specific flowers and birds) and to various human activities (including climbing a mountain, walking through city traffic, and placing the dead underground). Other composers have responded to more generalized features (such as sunsets, broad regions, and the world at war). Still others have derived their inspiration solely from their imaginations and, so, have represented imagined features or events. Some have been inspired by a people's songs and dances—folk activities. Despite the difficulty in representing ideas, places, activities in particular places, and so forth, in another medium, some composers have been inspired by poems, stories, legends, myths, and paintings and more to write their own representations of those works. Sometimes the resulting illustration is vivid; sometimes it is not. Therefore, the representations in music may be either crystal clear or, for want of a better word, mysterious.

Composers and geographers share a concern for such diverse physical geographical phenomena as mountains, seas, and the bleak Antarctic, among others. Interestingly, many composers have tried to represent physical processes in their music. Of course, as observed earlier, the composers are more concerned with the art of movement (e.g., of ocean waves, flowing rivers, soughing winds, diurnal and seasonal changes, speeding locomotives) than with the science of movement. Still, geographers and others surely can learn from the way composers capture or symbolize these processes in sound—those mysterious vibrations in air.

Equally, composers and geographers share an interest in how people—individuals and groups—live and interact within their respective national, regional, and local cultural landscapes and how they develop strong senses of and attachments to place. Composers' representations of people interacting in pastoral settings and cities, at festivals and dances, and much more, are numerous. Much of the action is pleasant, such as walking through a countryside, beside a lake, or in the bustle of a busy city, while other actions are violent, whether on a war front or in a storm.

A major message or implication in much of the music considered here pertains to the meaning of life. Composers, more than geographers, have explicitly expressed their personal concern for the human condition. For some, their music reflects an intensely personal search for meaning, while, for others, their music is a searing exposition on the harshness of life and death or on the joy of living. Some respond to the call for the heroic in a people's life, expressed in militaristic and nationalistic music; for others, there is an anger at lives lost or a wistful desire for past certainties in times of uncertainty. Life in the present (at the time of composing) is the major concern, but there is also an interest in the hereafter. Some music is delightfully, even painfully, reflective, while other music is bold and brash, devoid of explicit introspection.

Many composers clearly express a love of nature in their music. Gentle nature is often the focus of concern, whether via beautiful, intimate evocations of streams or in works that, more generally, include nature's sounds as part of performances. Raucous nature is a major theme, notably as represented in storms, at sea and over land, and volcanic eruptions. Some music is concerned with the challenge of uncompromising nature at its most dangerous. Creatures of the earth, such as birds, whales, and wolves, are represented in music, sometimes gracefully, sometimes humorously, sometimes threateningly. Some composers, tied to urban-industrial settings, seem remote from nature.

I have only discussed Ludwig van Beethoven, Tōru Takemitsu, Jean Sibelius, Alan Hovhaness, and, passingly, Gustav Mahler, Bedřich Smetana, and Richard Strauss, to any degree, in terms of their understanding of nature, as exhibited in some of their compositions. It was not my central purpose to develop a full appreciation of each composer's understandings of nature, as held at the times the many works considered were composed. Much more could be written about each of the seven mentioned composers and of the many others included in the book. This topic therefore remains as a subject for further study.

Due to space limitations, numerous works were omitted, principally because mention of them would have only redundantly illustrated the identified subthemes. For example, there is a large musical literature on representations of nature from around the time of Antonio Vivaldi and earlier, and I could have added numerous works by nineteenth-century romanticists and many by more recent composers. Clearly, further research should focus on individual subthemes, with greater representation of pertinent works, for a deeper understanding of the selected themes may then be achieved.

Geographers and composers may consider the same phenomena, yet their languages are so different that it may take some time and special effort before they are able to communicate. But each can learn from the other, as suggested, for instance, in the tone paintings of storms, rivers, geysers, volcanoes, landscapes being traversed, mythic tales, and so on. It was not my intention in this book to compare and contrast the approaches, but I do hope that any geographer and other readers will appreciate composers' insights and approaches that merit attention, for they can learn from the manner in which composers represent geographical features, processes, and events.

A word of (re)warning is needed: music is an abstraction, and while some music is representative of a story or scene or experience, the nature of the music's "message" may not be obvious to a listener upon first hearing; thus, a second hearing and background information may be needed before those landscapes in music can be appreciated for more than their abstract qualities. We must remember that Beethoven in his Symphony no. 6 did not include the birds, but representations of the birds; he did not include the actual brook, but representations of the brook; he did not include the actual scene of peasants, but representations of them. Further, Beethoven sought to evoke feelings about the landscape as much as he tried to represent the landscape itself. Also, significantly, his representation of the vivid summer day in a particular region, the people at play, the violent storm, and the return of serenity have not remained narrowly tied to place, for the music has taken on universal significance. These comments about Beethoven's marvelous symphony could be repeated for most of the works identified in the book.

The music identified here comes from several centuries of composition. A few of the works have been deposited in history's trash bin. Most of the works are still performed in concerts or are available on CDs. Whether old, or not so old, or, indeed, of very recent origin, the music can be listened to with benefit and enjoyment.

Much more might be written about the geography of music performance, indoors and outside. I hope that chapter 8 provides some useful suggestions concerning the type of research that is still required.

Finally, it is important at some point to simply sit back and enjoy what is being played (in a concert hall or on a radio, TV, or CD) without attempting to "dissect" it in order to search out this or that landscape reference. That said, knowing what lies behind a work can instill greater understanding about the work in question. As I mentioned at the outset, this book presents but one way to approach music: by considering how some composers have represented, structured, and symbolized real, imagined, and mythical landscapes in their tonal compositions. Applying the notions of landscapes in music can be rewarding to our understanding of many composers' intentions. Also, when we listen to the music, our lives can be enriched with beautiful sounds.

Appendix

On Soundscapes and the Geography of Music

People in many disciplines are interested in music, not just musicians, composers, conductors, listeners, and musicologists. These include scholars in anthropology, physics, psychology, religious studies and worship, sociology, native studies, and in a new field called biomusicology. In addition, researchers in the discipline of geography are interested in music.

Geographers are interested in variations over the surface of the earth of a wide variety of natural and human-made phenomena and processes of change, as well as their spatial distribution and spatial interaction, regionalization, and perceived meanings. Cultural geographers David Lowenthal (1975, 1976) and Susan Smith (1994) have nicely identified that sounds form part of the cultural landscape. And, as Smith (1994, p. 232) notes, "music, in particular, structures space and characterizes place." Whatever the reason for music, a remarkable diversity of musical traditions—involving distinctive melodies, harmonies, rhythms, instrumentation, nomenclature, and meanings—has been developed by different peoples who lived or live in varying locations of the vast surface of the earth (see Nash 1968, 1974, 1975a, 1975b). The array of musical traditions reflect peoples' various and contrasting cultural traditions and societal expectations and constraints. It is thus valid to talk of musical regions as a form of culture region, with the realization that there are many subregions. In general terms, we can identify very broad regions (each with a literature) as Europe, the Americas, Russia, Africa, South Asia, and Asia-Pacific. Influences from one region to the next, due to contact and the diffusion of ideas and material goods, have affected the way music has developed in different world regions. In the past half-century, musical styles, rhythms, and instrumentation increasingly have been intermingled to create a world musical style that incorporates all sorts of influences. Yet, there remain regional musical styles found in many parts of the world, the cultural cores and technical bases of which have existed for centuries (Nettl and Stone 1998–2002).

As noted in chapter 1, the term *soundscape* encompasses all environmental sounds around us, or as Canadian composer R. Murray Schafer (1977, p. 274) defines it, soundscape refers to "the sonic environment." Based on my conversations with David Lowenthal in the 1970s, I believe that he was the first geographer to use the term (Lowenthal 1975); however, as far as I can determine, Schafer concocted the term (1969, 1977, 1984, 1986), even though John Cage and other musicians in the United States (both before and then as contemporaries with Schafer) were exploring ways to make and use new sounds (see Rockwell 1997; Schwartz and Godfrey 1993). Schafer also concocted the term *acoustic ecology* and has published on the subject. Lowenthal participated in Schafer-inspired, interdisciplinary round table meetings in Canada on soundscapes in the mid-1970s and on acoustic ecology in the early 1990s. Lowenthal has a special interest in the "sonic" or "audible" past (1975; 1976; 1985, pp. 19–20). In addition to Lowenthal, the geographers to have provided thoughtful selective reviews on the concept of soundscapes are Douglas Pocock (1989), and especially Smith (1994). Other geographers, mostly in the United States, wrote in the early 1970s about various forms of music (e.g., Larry Ford and George Carney—see below), but the concept of soundscapes seems not to have caught their attention. Pocock (1987) of the Department of Geography at the University of Durham in England produced a fascinating fifty-minute tape entitled "Sound Portrait of a Cathedral City" (i.e., Durham), and Sara Cohen (1998) raises important questions about "sounding out the city." Geographer Ron Johnston's (1986) book on bell ringing in Britain is filled with valuable information.

Most research by geographers has been concerned with "vernacular" music. Vernacular (sometimes referred to as "popular") music in Western terms includes folk songs and dance music, ragtime, country music, jazz, rhythm and blues, rock-and-roll, pipe band music, and brass band music. Each of these can be studied from a variety of geographical perspectives. The important early explorative ruminations of Peter Nash (1968, 1974, 1975a, 1975b), Jean Tavenor (1970), and George Carney (1974, 1977a, 1977b, 1978, etc.; see below for additional examples) led the way for others to follow (Carney 2003b, pp. 1–11). Carney's project places music—but not orchestral music—centrally within folk geography.

Research directly on the geography of music has, to date, ranged wide (e.g., Violita 1980; Pocock 1988; Carney 1980b, 1990b, 1998a, 1998b, 2003a, pp. 1–11, 319–25; Kong 1995b; Nash and Carney 1996). The themes explored by geographers include (with some works cited more than once):

- *World music regions and subregions* (Nash 1968, 1975a, 1975b; Tavenor 1970)
- *National music regions and selected subregions* (Berland 1998; Carney 1980a, 1982, 1990a; Curtis and Rose 1983; Farrell 1998; Foley 2003; Gill 1993; Huefe 2003; Krim 1998; Sweeney-Turner 1998; Thirwall 1992)
- *Issues of local and national identity and images of place* (Foley 2003; Gumprecht 1998; Lehr 1983, 1985; Stradling 1998; Wall 2000; Leonard 2005)
- *Personalities* (Carney 1999b, 2001; Cohen 1998; Curtis 1976; Flynn 2003; Gold 1998a, 1998b; Krim 2003; Kuhlken 2003; Revill 1995, 1998)
- *Perceptions and sense of place* (Ford and Henderson 1974; Lehr 1983; Kracht 1989; Stradling 1998)
- *Environments* (Nash 1975b; Henderson 1978; Woods and Gritzner 1982)
- *Use of radio stations* (Carney 1974, 1977b)
- *Locations of origin and routes of diffusion of selected styles of music, namely*
    - *blues* (Arkell 1991; Carney 2003d)
    - *jazz* (Glasgow 1979; Stump 1998)
    - *soul* (Hollows and Milestone 1998)
    - *rock and roll, and rock* (Bell 1998; Butler 1984; Carney 1999a; Dixon 1988; Ford 1971; Francaviglia 1973; Gordon 1970; McLeay 1994, 1995)
    - *bluegrass* (Carney 1974)
    - *country and western* (Carney 1977a, 1977b, 1980a, 1987, 1994a, 2000; Gritzner 1978; Lehr 1983, 1985; D. Meyer 1976)
    - *fiddling* (Crowley 1984)
    - *brass bands* (Herbert 1998)
    - *folk* (Carney 1990a; Kracht 1989; Revill 2004, in press)
    - *pop* (Paterson 1991)
    - *rave* (Halfacree and Kitchin 1996; Gibson 1999)
    - *rap* (Wall 2000; Carney 2003c)
- *Indigenous peoples* (Gibson 1998; Gibson and Dunbar-Hall 2000)
- *Locations and associated landscapes referred to in music and songs* (Cohen 1991, 1995; Lehr 1983; Rycroft 1998)
- *Buildings* (P. Jackson 1989, pp. 86–87; Woolf 1988)
- *Carnival* (P. Jackson 1989)

- *Tourism* (Gibson and Connell 2003)
- *Music and racism* (P. Jackson 1989; Wall 2000)
- *The relationships between the local and the global* (Kong 1995a; Lovering 1998; Smith 1994; Connell and Gibson 2003)

A useful map and text presentation of some of the research to the late 1980s appears in Carney's chapter on music and dance in *This Remarkable Continent: An Atlas of the United States and Canadian Society and Cultures* (1982). Carney has also produced four edited collections, *The Sounds of People and Places: Readings in the Geography of Music* (1978), *The Sounds of People and Places: Readings in the Geography of American Folk and Popular Music* (1987), *The Sounds of People and Places: A Geography of American Folk and Popular Music* (1994b), and *The Sounds of People and Places: A Geography of American Music from Country to Classical and Blues to Bop* (2003a). A useful collection of essays appears in *Musikgeographie* (Büttner, Schnabel, and Winkler 1990), and a new collection of original essays appears in *Geography and Music* (Waterman and Brunn, in press).

A recent basic cultural-geography text includes two chapters that directly deal with music, a refreshing development (Cook et al. 2000). An edited book collection of essays entitled *The Place of Music* (Leyshon, Matless, and Revill 1998; which takes from Leyshon, Matless, and Revill 1995) is by British writers who, William Norton (2000, pp. 304–5) has observed, have pointedly ignored the considerable literature by North American geographers. The book explores some of the above-noted themes, as well as the globalization of music production and marketing, alternative and hybridized music scenes as sites of localized resistance, the nature of soundscapes, and issues of migration and national identity. Of this book, Susan Smith (1994) observes that the sixteen authors provide an "engaging tour through the political economy of sound and the cultural politics of sound to the aesthetics of listening and the poetics of performance" (comment on back cover).

John Connell and Chris Gibson's *Sound Tracks: Popular Music, Identity, and Place* (2003) is a major geographical exploration of popular music, in which music and the music industry are situated within spatial theories of globalization and local change. *Music, Space and Place: Popular Music and Cultural Identity* (2004), edited by Sheila Whiteley, Andy Bennett, and Stan Hawkins, examines the urban and rural spaces in which popular music is experienced, produced, and consumed. Though not written by geographers, the book nevertheless offers interesting geographical insights.

Though it pertains to an instrument and not to music as such, note should be made of an important map showing regions for seven types of dulcimer and possible routes of diffusion. The core regions identified are Europe, West Asia, and East Asia (Kettlewell 2001).

The principal theme "missing" from most writings by geographers pertains to orchestras (Knight, in press) and orchestral music. George Revill stands out as a notable exception due to his two fine articles on pastoralism in music, which raise parallels between the study of music and the study of landscapes (1991, 2000). Other than Revill's work, Nash's early studies, noted above, Rolf Sternberg's (1998) exploration of aspects of Wagner's operas, and John Flynn's (2003) focus on singers at the Metropolitan Opera in New York, geographers have tended to shy away from classical music, though the following can be noted: Stanley Waterman's (1998a, 1998b) concern for classical music festivals in relation to societal coherence and national identities; Robert Stradling's (1998) consideration of "sensibilities of place" in English music; and Revill's two (1995, 1998) biographical comments on the English composer Samuel Coleridge-Taylor. Clearly, the present volume is a marked departure from the works written by geographers to this time.

Finally, cultural geographers have traditionally organized their thoughts around certain basic themes, some of which are identified above. The themes are culture areas or cultural regions (the study of the spread of universals and traits over the surface of the earth, which can lead to the identification of regional distributions), culture history (the study of how some artifact—or a whole culture—came to be, notably with concern for where it originated and then by what routes and under what circumstances it was dispersed to other places), cultural ecology (or the dynamics of human-environment interaction), and cultural landscape (the landscape as molded and modified by human action). As Phillip Wagner and Marvin Mikesell remarked many years ago, to focus on any one of these themes automatically involves consideration of each of the others (1962, p. 22) and, ultimately, reveals basic qualities about a culture. A case can be made for giving primacy to each one of these themes, but the one chosen for this book is landscape. The concept of cultural landscape is interactive, involving human:nature interaction, that is, human (cultural) groups and nature in its fullest sense. Implicit in the use of the colon is the perception of the landscape, for any cultural landscape is "seen" and "appreciated," and ultimately "represented," by humans through the prism of personal, societal, and cultural "lenses" or, in other words, perception as influenced by both personal and broader societal and cultural attitudes and values. Also implicit here is the issue of control—control of individuals and groups by society. Such controls can impinge upon composers, as identified in several chapters.

One further observation: Mention must be made of the geographical and musical Silk Road Project (www.silkroadproject.org) initiated by internationally renowned cellist Yo-Yo Ma, the project's artistic director. Over the course of his performing career and travels, Ma became increasingly intrigued by the migration of ideas, instruments, and performance styles among communities along the historic trade route that connected the peoples and traditions of Asia with those of Europe. The best known of these routes went from the ancient Chinese capital of Chang'an (Xian), diverged into northern and southern routes that skirted the Central Asian Taklamakan Desert, converged to cross the Iranian plateau, and continued on to the eastern shores of the Mediterranean and to cities like Antioch and Tyre. Ma's fascination with what he discovered led him, in 1998, to found the project to study the "ebb and flow of ideas" in those areas today. The project acts as an umbrella organization and common resource for a number of artistic, cultural, and educational programs, drawing, as they do, artists from many countries who share their living traditions and musical voices in recordings and concerts.

References

I have not listed here hundreds of books and journal articles that I consulted but did not directly quote or cite; nor have I listed numerous full (or study) scores that I examined for this study.

Abell, Arthur M. 1994. *Talks with Great Composers.* New York: Citadel Press.

Amis, John, and Michael Rose. 1992. *Words about Music.* New York: Paragon House.

Anderson, Keith. 2001. "Granville Bantock." Pp. 2–3 in Bantock, *Hebridean Symphony,* CD–NAXOS 8.555473.

Archer, Thomas. 1968. "Champagne, Claude." *The Canadian Music Journal,* vol. 2, no. 2, pp. 3–10.

Arkell, Thomas. 1991. "Geography on Record: Origins and Diffusion of the Blues." *Geographical Magazine,* vol. 63, pp. 30–34.

Ashley, Tim. 1999. *Richard Strauss.* London: Phaidon.

Austin, William W. 1986. "Aaron Copland." *The New Grove Dictionary of American Music,* vol. 1, ed. H. Wiley Hitchcock and Stanley Sadie. London: Macmillan.

Barenboim, Daniel, and Edward W. Said. 2004. *Parallels and Paradoxes: Explorations in Music and Society,* ed. Ara Guzelman. New York: Vintage Books.

Barford, Philip. 1978. *Bruckner Symphonies.* London: Ariel Music, BBC Publications.

Barth, Karl. 1986. *Wolfgang Amadeus Mozart.* Grand Rapids, MI: Eerdmans.

Bartoš, František. 1951. *Mà Vlast: Cyklus Symfonickych Básní.* Praha: Editio Supraphon.

Bell, Thomas L. 1998. "Why Seattle: An Examination of an Alternative Rock Culture Hearth." *Journal of Cultural Geography,* vol. 18, pp. 35–47.

Berland, Jody. 1998. "Locating Listening: Technological Space, Popular Music, and Canadian Mediations." Pp. 129–50 in *The Place of Music,* ed. Andrew Leyshon, David Matless, and George Revill. New York: Guilford Press.

Berlioz, Hector. [1862] 2000. *A Critical Study of Beethoven's Nine Symphonies*, trans. Edwin Evans. Urbana: University of Illinois Press.

Bernbaum, Edwin. 1997. *Sacred Mountains of the World*. Berkeley: University of California Press.

Berndt, Ronald M., and C. H. Berndt. 1964. *The World of the First Australians*. Chicago: University of Chicago Press.

Bernstein, Leonard. 1959. *The Joy of Music*. New York: Simon and Schuster.

Botstein, Leon. 1991. "The Aesthetics of Assimilation and Affirmation: Reconstructing the Career of Felix Mendelssohn." Pp. 5–42 in *Mendelssohn and His World*, ed. R. L. Todd. Princeton, NJ: Princeton University Press.

Bowen, Meiron, ed. 1995. *Tippett on Music*. Oxford, UK: Clarendon Press.

Boyd, Malcolm. 2001. "National Anthems." Pp. 654–55 in *The New Grove Dictionary of Music and Musicians*, vol. 17, ed. Stanley Sadie. 2nd ed. New York: Macmillan.

Brant, Harry. 1998. "Space as an Essential Aspect of Musical Composition." Pp. 223–42 in *Contemporary Composers on Contemporary Music*, ed. Elliot Schwartz and Barney Childs. New York: Da Capo Press.

Breuer, János. 1990. *A Guide to Kodály*. Budapest: Corvina.

Britten, Benjamin. 1962. *War Requiem*. London: Boosey and Hawkes.

Brockman, Norbert C. 1997. *Encyclopedia of Sacred Places*. New York: Oxford University Press.

Brown, Ralph H. 1948. *Historical Geography of the United States*. New York: Harcourt Brace.

Broyles, Michael. 1986. "Ensemble Music Moves out of the Private House." Pp. 101–26 in *The Orchestra: Origins and Transformations*, ed. Joan Peyser. New York: Billboard Books.

Burke, Albert. 1956. "Influence of Man upon Nature—The Russian View: A Case Study." Pp. 1035–51 in *Man's Role in Changing the Face of the Earth*, ed. William L. Thomas Jr. Chicago: University of Chicago Press.

Burke, John. 1983. *Musical Landscapes*. Exeter, UK: Webb & Bower.

Butler, Richard W. 1984. "The Geography of Rock: 1954–1970." *Ontario Geographer*, no. 24, pp. 1–33.

Büttner, Manfred, Wolfgang Schnabel, and Klaus Winkler, eds. 1990. *Musikgeographie*. Bochum, Germany: Universitätsverlag.

Campbell, Joseph. 1968. *The Masks of God: Creative Mythology*. New York: Viking.

Carney, George O. 1974. "Bluegrass Grows All Around: The Spatial Dimensions of a Country Music Style." *Journal of Geography*, vol. 73, pp. 34–55.

———. 1977a. "From Down Home to Uptown: The Diffusion of Country Music Radio Stations in the United States." *Journal of Geography*, vol. 75, pp. 104–10.

———. 1977b. "Spatial Diffusion of the All-Country Music Radio Stations in the United States, 1971–1974." *John Edwards Memorial Foundation Quarterly*, vol. 13, pp. 58–66.

———, ed. 1978. *The Sounds of People and Places: Readings in the Geography of Music*. Washington, DC: University Press of America.

———. 1980a. "Country Music and the South: A Cultural Geography Perspective." *Journal of Cultural Geography*, vol. 1, pp. 16–33.

———. 1980b. "Geography of Music: A Bibliography." *Journal of Cultural Geography*, vol. 1, pp. 185–86.

———. 1982. "Music and Dance." Pp. 234–53 in *This Remarkable Continent: An Atlas of the United States and Canadian Society and Cultures*, ed. John F. Rooney Jr., Wilbur Zelinsky, and Dean R. Louder. College Station: Texas A&M University Press.

———. 1987. "Okie from Muskogee: The Spatial Dimensions of Oklahoma Country Music." *Arkansas Journal of Geography*, vol. 3, pp. 11–17.

———. 1990a. "The Ozarks: A Reinterpretation Based on Folk Song Materials." *North American Culture*, vol. 6, pp. 40–55.

———. 1990b. "The Geography of Music: Inventory and Prospect." *Journal of Cultural Geography*, vol. 10, pp. 35–48.

———. 1994a. "Branson: The New Mecca of Country Music." *Journal of Cultural Geography*, vol. 14, 17–32.

———, ed. 1994b. *The Sounds of People and Places: A Geography of American Folk and Popular Music*. 3rd ed. Lanham, MD: Rowman & Littlefield.

———. 1998a. "Music Geography." *Journal of Cultural Geography*, vol. 18, no. 1, pp. 1–10.

———, ed. 1998b. *Journal of Cultural Geography*, vol. 18, no. 1, pp. 1–97. Special issue with six essays on the geography of music.

———. 1999a. "Cowabunga! Surfer Rock and the Five Themes of Geography." *Popular Music and Society*, vol. 23, pp. 3–29.

———. 1999b. *From Lee to Reba: Oklahoma Women in American Popular Music*. Stillwater: Oklahoma University Press.

———. 2000. "Western Swing in Fort Worth: Culture Hearth of the First Alternative Country Music Form." *Southwestern Geographer*, vol. 4, pp. 1–15.

———. 2001. "Rockin' and Rappin' in American Music: Themes and Resources." *Journal of Geography*, vol. 100, pp. 261–70.

———, ed. 2003a. *The Sounds of People and Places: A Geography of American Music from Country to Classical and Blues to Bop*. 4th ed. Lanham, MD: Rowman & Littlefield.

———. 2003b. Introduction to *The Sounds of People and Places: A Geography of American Music from Country to Classical and Blues to Bop*, ed. George O. Carney. 4th ed. Lanham, MD: Rowman & Littlefield.

———. 2003c. "Rappin' in America: A Regional Music Phenomenon." Pp. 93–118 in *The Sounds of People and Places: A Geography of American Music from Country to Classical and Blues to Bop*, ed. George O. Carney. 4th ed. Lanham, MD: Rowman & Littlefield.

———. 2003d. "Urban Blues: The Sound of the Windy City." Pp. 241–54 in *The Sounds of People and Places: A Geography of American Music from Country to Classical and Blues to Bop*, ed. George O. Carney. 4th ed. Lanham, MD: Rowman & Littlefield.

Chase, Gilbert. 1966. *America's Music*. New York: McGraw-Hill.

Chou, Wen-chung. 1998. "Towards a Re-Merger in Music." Pp. 309–15 in *Contemporary Composers on Contemporary Music*, ed. Elliot Schwartz and Barney Childs. New York: Da Capo Press.

Cohen, Sara. 1991. "Popular Music and Urban Regeneration: The Music Industries of Merseyside." *Cultural Studies*, vol. 5, pp. 332–46.

———. 1995. "Sounding Out the City: Music and the Sensuous Production of Place." *Transactions of the Institute of British Geographers*, vol. NS20, pp. 434–46.

——. 1998. "Sounding Out the City: Music and the Sensuous Production of Place." Pp. 269–90 in *The Place of Music*, ed. Andrew Leyshon, David Matless, and George Revill. New York: Guilford Press.
Connell, John, and Chris Gibson. 2003. *Sound Tracks: Popular Music, Identity, and Place*. New York: Routledge.
Conrad, Ulrich. 2001. "Architecture as a Musical Instrument." Available at www.berlin-philharmonic.com/engl/6phil/b60101c.htm (accessed November 6, 2001).
Cook, Ian, et al. 2000. *Cultural Turns: Geographical Turns*. Harlow, Essex: Prentice Hall.
Cooney, Denise von Glahn. 1995. *Reconciliations: Time, Space, and the American Place in the Music of Charles Ives*. PhD diss., University of Washington.
Copland, Aaron. [1939] 1988. *What to Listen for in Music*. New York: Penguin.
Cosgrove, Denis, and Stephen Daniels, eds. 1988. *The Iconography of Landscape*. Cambridge: Cambridge University Press.
Cousins, Norman. 1979. *Anatomy of an Illness as Perceived by the Patient*. New York: Norton.
Cresswell, Lyell. 2001. "Dancing on a Volcano." P. 9 in *Landscapes: New Zealand Orchestral Music*. Wellington: Morrison Music Trust, CD–MMT2037.
Crory, Neil. 1996. "Takemitsu and Beethoven." Pp. 36–38 in *Performance*. Toronto, ON: Toronto Symphony Orchestra, June 18–22.
Crowley, John M. 1984. "Old-Time Fiddling in Big Sky Country." *Journal of Cultural Geography*, vol. 5, pp. 47–60.
Culshaw, John. 1963. Commentary to Benjamin Britten's *War Requiem* (London Records A4255), p. 5.
Curtis, James R. 1976. "Woody Guthrie and the Dust Bowl." *Places*, vol. 3, pp.1–14.
Curtis, James R., and Richard F. Rose. 1983. "'The Miami Sound': A Contemporary Latin Form of Place-Specific Music." *Journal of Cultural Geography*, vol. 4, pp. 110–18.
Darnton, Christian. 1940. *You and Music*. London: Pelican Books.
Dart, William. 1995. "New Zealand Orchestral Music." *New Zealand Composers*, CCD1073.
Davie, Cedric Thorpe. 1980. *Scotland's Music*. Oxford, UK: Blackwood.
Davies, Stephen. 1994. *Musical Meaning and Expression*. Ithaca, NY: Cornell University Press.
de Blij, Harm J. 1983. *Wine: A Geographic Appreciation*. Totowa, NJ: Rowman & Allenheld.
——. 1985. *Wine Regions of the Southern Hemisphere*. Totowa, NJ: Rowman & Allenheld.
——. 2000. *Wartime Encounter with Geography*. Lewes, Sussex: Book Guild.
Debussy, Claude. [n.d.] 1997. *La Mer*. New York: Dover.
Dixon, Bruce. 1988. *Geography of Rock Music*. Waterloo, ON: University of Waterloo.
Downes, Edward. 1976. *Guide to Symphonic Music*. New York: Walker.
Downes, Olin. 1943. "Arthur Honegger: Pacific 231." Pp. 232–33 in *The Music Lover's Handbook*, ed. Elie Siegmeister. New York: William Morrow.

Downs, Roger M., and David Stea. 1977. *Maps in Mind*. New York: Harper and Row.
Druskin, Mikhail. 1983. *Igor Stravinsky*, trans. Martin Cooper. Cambridge: Cambridge University Press.
Dubal, David. 2001. *The Essential Canon of Classical Music*. New York: North Point Press.
Dyer, Richard. 1986. *Ellen Taaffe Zwilich: Symphony No. 1, Celebration, Prologue and Variations*. New York: New World Records, NW336-2.
Eliade, Mircea. 1961. *The Sacred and the Profane*. New York: Harper Torchbook.
Ellman, Donald. 1995. "The Symphony in Nineteenth-Century Germany." Pp. 351–62 in *A Guide to the Symphony*, ed. Robert Layton. Oxford, UK: Oxford University Press.
Engel, Carl. 1866. *An Introduction to the Study of National Music*. London: Longmans, Green, Reader, and Dyer.
Evernden, Neil. 1992. *The Social Creation of Nature*. Baltimore: Johns Hopkins University Press.
Fanning, David J. 1995. "Nielsen." Pp. 351–62 in *A Guide to the Symphony*, ed. Robert Layton. Oxford, UK: Oxford University Press.
Farrell, Gerry. 1998. "The Early Days of the Gramophone Industry in India." Pp. 57–82 in *The Place of Music*, ed. Andrew Leyshon, David Matless, and George Revill. New York: Guilford Press.
Feder, Stuart. 1992. *Charles Ives: "My Father's Song."* New Haven, CT: Yale University Press.
Finkelstein, Sydney. 1960. *Composer and Nation: Folk Heritage in Music*. New York: International Publishers.
Flynn, John J. 2003. "Americans at the Met: The Rise of the Homegrown Opera Star in the Twentieth Century." Pp. 149–63 in *The Sounds of People and Places: A Geography of American Music from Country to Classical and Blues to Bop*, ed. George O. Carney. 4th ed. Lanham, MD: Rowman & Littlefield.
Foley, Hugh W., Jr. 2003. "From Stomp Dances to Dew Songs: American Indian Music in Oklahoma at the Turn of the Twenty-first Century." Pp. 27–49 in *The Sounds of People and Places: A Geography of American Music from Country to Classical and Blues to Bop*, ed. George O. Carney. 4th ed. Lanham, MD: Rowman & Littlefield.
Foote, Kenneth E. 1997. *Shadowed Ground*. Austin: University of Texas Press.
Foote, Kenneth E., Peter J. Hugill, and Kent Mathewson, eds. 1994. *Re-Reading Cultural Geography*. Austin: University of Texas Press.
Ford, Larry R. 1971. "Geographic Factors in the Origins, Evolution and Diffusion of Rock and Roll." *Journal of Geography*, vol. 70, pp. 455–64.
Ford, Larry R., and Floyd M. Henderson. 1974. "The Image of Place in American Popular Music: 1890–1970." *Places*, no. 1, pp. 31–37.
Four Days: The Historical Record of the Death of President Kennedy. 1963. New York: American Heritage.
Francaviglia, R. V. 1973. "Diffusion and Popular Culture: Comments on the Spatial Aspects of Rock Music." Pp. 87–96 in *An Invitation to Geography*, ed. David A. Lanegran and R. Palm. New York: McGraw-Hill.

Fuller, Sophie. 1994. *The Pandora Guide to Women Composers: Britain and the United States, 1629–Present*. London: Pandora.

Gibson, Chris. 1998. "'We Sing Our Home, We Dance Our Land': Indigenous Self-Determination and Contemporary Geopolitics in Australian Popular Music." *Environment and Planning D*, vol. 16, pp. 163–84.

———. 1999. "Subversive Sites: Rave Culture, Spatial Politics and the Internet in Sydney, Australia." *Area*, vol. 31, pp. 19–33.

Gibson, Chris, and John Connell. 2003. "'Bongo Fury': Tourism, Music and Cultural Economy at Byron Bay, Australia." *Tijdschrift voor Economische en Sociale Geografie*, vol. 94, no. 2, pp. 164–87.

Gibson, Chris, and P. Dunbar-Hall. 2000. "'Nitmiluk': Place and Empowerment in Aboriginal Popular Music." *Ethnomusicology*, vol. 44, pp. 39–64.

Gill, Warren. 1993. "Region, Agency, and Popular Music: The Northwest Sound, 1958–1966." *Canadian Geographer*, vol. 37, no. 2, 120–31.

Gillies, Malcolm, and David Pear. 2001. "Grainger, Percy." Pp. 269–73 in *The New Grove Dictionary of Music and Musicians*, vol. 10, ed. Stanley Sadie. 2nd ed. New York: Macmillan.

Gilman, Lawrence. 1966. *Nature in Music and Other Studies in the Tone-Poetry of Today*. Freeport, NY: Books for Libraries Press.

Glacken, Clarence. 1967. *Traces on the Rhodian Shore*. Berkeley: University of California Press.

Glasgow, Jon A. 1979. "An Example of Spatial Diffusion: Jazz Music." *Geographical Survey*, vol. 8, pp. 10–21.

Gold, John R. 1998a. "From 'Dust Storm Disaster' to 'Pastures of Plenty': Woody Guthrie and Landscapes of the American Depression." Pp. 249–68 in *The Place of Music*, ed. Andrew Leyshon, David Matless, and George Revill. New York: Guilford Press.

———. 1998b. "Roll on Columbia: Woody Guthrie, Migrants' Tales, and Regional Transformation in the Pacific Northwest." *Journal of Cultural Geography*, vol. 18, pp. 83–97.

Gordon, Jeffrey J. 1970. *Rock-and-Roll Music: A Diffusion Study*. Unpublished master's thesis, Pennsylvania State Univ.

Gottmann, Jean. 1973. *The Significance of Territory*. Charlottesville: The University Press of Virginia.

Graber, Linda H. 1976. *Wilderness as Sacred Space*. Monograph Series no. 8. Washington, DC: Association of American Geographers.

Gray, Alice. 1995. Introduction to *Ansel Adams: The National Park Service Photographs*. New York: Artabras, Abbeville Publishing.

Gritzner, Charles F. 1978. "Country Music: A Reflection of Popular Culture." *Journal of Popular Culture*, vol. 11, pp. 857–64.

Grout, Donald Jay, and Claude V. Palisca. 1996. *A History of Western Music*, 5th ed. New York: Norton.

Gumprecht, Blake. 1998. "Lubbock on Everything: The Evocation of Place in the Music of West Texas." *Journal of Cultural Geography*, vol. 18, pp. 61–81.

Gup, Ted. 2000. "Empire of the Dead." *Smithsonian*, vol. 31, no. 1, pp. 106–13. Reprinted in 2001 as "Empire of the Dead." *Doctor's Review*, vol. 19, no. 4, pp. 140–45 (some of the illustrations differ).

Halfacree, Keith H., and Robert M. Kitchin. 1996. "'Madchester Rave On': Placing the Fragments of Popular Music." *Area*, vol. 28, pp. 47–55.
Hall, Michael. 1996. *Leaving Home: A Conducted Tour of Twentieth-Century Music with Simon Rattle*. London: Faber and Faber.
Hallmark, Rufus. 1986. "The Star Conductor and Musical Virtuosity." Pp. 545–76 in *The Orchestra: Origins and Transformations*, ed. Joan Peyser. New York: Charles Scribner.
Hanslick, Eduard. 1963. *Music Criticisms 1846–99*, trans. and ed. Henry Pleasants. Harmondsworth, UK: Penguin.
Hanson, Lawrence, and Elizabeth Hanson. 1965. *Tchaikovsky*. London: Cassell.
Harley, Maria. 1993. "From Point to Sphere: Spatial Organization of Sound in Contemporary Music (after 1950)." *Canadian University Music Review*, vol. 13, pp. 123–44.
Haskell, Harry. 1996. *The Attentive Listener: Three Centuries of Music Criticism*. London: Faber and Faber.
Henderson, Floyd. 1978. "The Image of New York City in American Popular Music: 1890–1970." *New York Folk Quarterly*, vol. 11, pp. 267–79.
Hepokoski, James. 1993. *Sibelius: Symphony No. 5*. Cambridge: Cambridge University Press.
Herb, Guntrum H., and David H. Kaplan, eds. 1999. *Nested Identities*. Lanham, MD: Rowman & Littlefield.
Herbert, Trevor. 1998. "Victorian Brass Bands: Class, Taste, and Space." Pp. 104–28 in *The Place of Music*, ed. Andrew Leyshon, David Matless, and George Revill. New York: Guilford Press.
Herresthal, Harald. 2004. *Edvard Grieg*. Oslo: Ministry of Foreign Affairs.
Hewitt, Kenneth. 1997a. "Place Annihilation: Air War and the Vulnerability of Cities." Pp. 296–320 in *Regions at Risk: Geographical Introduction to Disasters*. London: Longman.
———. 1997b. "The Holocaust: Genocide and Geographical Calamity." Pp. 321–48 in *Regions at Risk: Geographical Introduction to Disasters*. London: Longman.
Hickok, Robert. 1993. *Exploring Music*. 5th ed. Dubuque, IA: WCB Brown & Benchmark.
Hill, Ralph. 1947. "Sibelius the Man." Pp. 9–13 in *The Music of Sibelius*, ed. Gerald Abraham. New York: Norton.
Hollows, Joanne, and Katie Milestone. 1998. "Welcome to Dreamsville: A History and Geography of Northern Soul." Pp. 83–103 in *The Place of Music*, ed. Andrew Leyshon, David Matless, and George Revill. New York: Guilford Press.
Honegger, Arthur. [1923] 1975. *Pacific 231: Mouvement symphonique*. Paris: Editions Salabert.
Huefe, Edward G., III. 2003. "Music Geography across the Borderline: Musical Iconography, Mythic Themes, and North American Place Perceptions of the United States–Mexico Borderlands." Pp. 51–77 in *The Sounds of People and Places: A Geography of American Music from Country to Classical and Blues to Bop*, ed. George O. Carney. 4th ed. Lanham, MD: Rowman & Littlefield.
Isaac, Eric. 1964–1965. "God's Acre." *Landscape*, vol. 14, no. 2, pp. 28–32.
Ives, Charles. 1972. *Memos*, ed. John Kirkpatrick. New York: Norton.

Jackson, John B. 1984. *Discovering the Vernacular Landscape*. New Haven, CT: Yale University Press.

Jackson, Peter. 1989. *Maps of Meaning*. London: Unwin Hyman.

Jezic, Diane Peacock, with Elizabeth Wood. 1988. *Women Composers*. 2nd ed. New York: Feminist Press at the City University of New York.

Johnson, Robert Sherlaw. 1975. *Messiaen*. Berkeley: University of California Press.

Johnston, Ronald J. 1986. *Bell-Ringing: The English Art of Change Ringing*. London: Viking Press.

Joyce, Lionel. 2001. "Music: The Great Healer." *BBC Music Magazine*, June, p. 6.

Kastendieck, Miles. 1964. "Hovhaness, Alan." P. 1008 in *The International Cyclopedia of Music and Musicians*, ed. Oscar Thompson. New York: Dodd, Mead and Co.

Kennedy, Michael. 1984. *Strauss Tone Poems*. London: BBC Publications.

———. 2000. *Mahler*. Oxford, UK: Oxford University Press.

Kettlewell, Robert. 2001. "Map Showing the Distribution of the Dulcimer and Possible Routes of Dissemination." P. 684 in *The New Grove Dictionary of Music and Musicians*, vol. 7, ed. Stanley Sadie. 2nd ed. London: Macmillan.

Khan, Hazrat Inayat. 1996. *The Mysticism of Sound and Music*. Boston: Shambhala.

Kivy, Peter. 1990. *Music Alone: Philosophical Reflection on the Purely Musical Experience*. Ithaca, NY: Cornell University Press.

Knight, David B. 1982a. *Cemeteries as Living Landscapes*. Ottawa: Ontario Genealogical Society.

———. 1982b. "Identity and Territory." *Annals of the Association of American Geographers*, vol. 72, pp. 514–31.

———. 1983. "Urban Images and Cognitive Mapping." *Ontario Geographer*, vol. 20, pp. 29–48.

———. 1986. "Perceptions of Landscapes in Heaven." *Journal of Cultural Geography*, vol. 6, pp. 127–38.

———. 1987. "The Other Side of the Tracks: Perceptions of an Urban Place." *Journal of Geography*, vol. 86, pp. 14–18.

———. 1991. *Choosing Canada's Capital*. Ottawa, ON: Carleton University Press.

———. 1998. "Extending the Local: The Small Town and Globalization." *GeoJournal*, vol. 45, no. 1, pp. 145–49.

———. 1999a. "Afterword: Nested Identities, Nationalism, Territory, and Scale." Pp. 317–29 in *Nested Identities*, ed. Guntram H. Herb and David H. Kaplan. Lanham, MD: Rowman & Littlefield.

———. 1999b. "People Together, Yet Apart: Rethinking Territory, Sovereignty, and Identities." Pp. 209–26 in *Reordering the World*, ed. George Demko and W. Wood. 2nd ed. Boulder, CO: Westview.

———. 2005. "Cemeteries." *The Canadian Encyclopedia*, ed. James H. Marsh. Toronto: Historica Foundation of Canada, available at www.thecanadianencyclopedia.com (accessed September 7, 2005).

———. In press. "Geographies of the Orchestra." In *Geography and Music*, ed. Stanley Waterman and Stanley D. Brunn. Dordrecht: Kluwer-Springer.

Kolodin, Irving, et al. 2001. "New York: 5. Orchestras and Bands." Pp. 827–29 in *The New Grove Dictionary of Music and Musicians*, vol. 17, ed. Stanley Sadie. 2nd ed. New York: Macmillan.

Kong, Lily. 1995a. "Music and Cultural Politics: Ideology and Resistance in Singapore." *Transactions of the Institute of British Geographers*, vol. 20, pp. 447–59.

———. 1995b. "Popular Music in Geographical Analysis." *Progress in Human Geography*, vol. 19, no. 2, pp. 183–98.

Kracht, James B. 1989. "Perception of the Great Plains in Nineteenth-Century Folk Songs: Teaching about Place." *Journal of Geography*, vol. 88, pp. 206–12.

Kramer, Jonathan D. 1988. *Listen to the Music*. New York: Schirmer Books, Macmillan.

Krim, Arthur. 1998. "'Get Your Kicks on Route 66!': A Song Map of Postwar Migration." *Journal of Cultural Geography*, vol. 18, pp. 49–60.

———. 2003. "Stephen Foster and His Musical Geography of the South." Pp. 15–25 in *The Sounds of People and Places: A Geography of American Music from Country to Classical and Blues to Bop*, ed. George O. Carney. 4th ed. Lanham, MD: Rowman & Littlefield.

Kuhlken, Robert T. 2003. "Louie Louie Land: Music Geography of the Pacific Northwest." Pp. 277–312 in *The Sounds of People and Places: A Geography of American Music from Country to Classical and Blues to Bop*, ed. George O. Carney. 4th ed. Lanham, MD: Rowman & Littlefield.

Küng, Hans. 1992. *Traces of Transcendence*. London: SCM Press.

Kupferberg, Herbert. 1976. *Tanglewood*. New York: McGraw-Hill.

Lawson, Colin, and Robin Stowell. 1999. *The Historical Performance of Music*. Cambridge, U.K.: Cambridge University Press.

Layton, Robert. 1983. *Sibelius*. London: Dent.

Lee, Joanna C. 2001. "Chou Wen-chung." Pp. 789–90 in *The New Grove Dictionary of Music and Musicians*, vol. 5, ed. Stanley Sadie. 2nd ed. London: Macmillan.

Lehr, John C. 1983. "Texas (When I Die): National Identity and Images of Place in Canadian Country Music Broadcasts." *Canadian Geographer*, vol. 27, pp. 361–70.

———. 1985. "As Canadian as Possible . . . under the Circumstances: Regional Myth, Images of Place, and National Unity in Canadian Country Music." *Journal of Country Music*, vol. 2, pp. 16–19.

Leonard, Marion. 2005. "Performing Identities: Music and Dance in the Irish Communities of Coventry and Liverpool." *Social and Cultural Geography*, vol. 6, no. 4, pp. 515–29.

Leyshon, Andrew, David Matless, and George Revill. 1995. "The Place of Music." *Transactions of the Institute of British Geographers*, vol. NS20, no. 4, pp. 423–33.

———, eds. 1998. *The Place of Music*. New York: Guilford Press.

Lovering, John. 1998. "The Global Music Industry." Pp. 31–56 in *The Place of Music*, ed. Andrew Leyshon, David Matless, and George Revill. New York: Guilford Press.

Lowenthal, David. 1975. "The Audible Past." Pp. 209–17 in *The Canadian Music Book*, vols. 11 and 12. Montreal: International Music Council.

———. 1976. "Turning into the Past: Can We Recapture the Soundscapes of Bygone Days?" *UNESCO Courier*, vol. 29, pp. 15–21.

———. 1985. *The Past Is a Foreign Country*. Cambridge: Cambridge University Press.

Lowenthal, David, and Martyn J. Bowden. 1976. *Geographies of the Mind: Essays in Historical Geosophy*. New York: Oxford University Press.

Maconie, Robin. 1993. *The Concept of Music*. Oxford, UK: Clarendon.

March, Ivan, Edward Greenfield, and Robert Layton. 1999. *The Penguin Guide to Compact Discs*. London: Penguin.

Marett, Allan, Catherine J. Ellis, and Margaret Gummow. 2001. "Australia, Aboriginal Music." Pp. 193–214 in *The New Grove Dictionary of Music and Musicians*, vol. 2, ed. Stanley Sadie. 2nd ed. New York: Macmillan.

Marx, A. B. 1865. *Erinnergungen aus mainen Leben*, 2 vols. [Berlin], translated in part in 1991 as *Mendelssohn and His World*, trans. R. L. Todd. Princeton, NJ: Princeton University Press.

Mather, Bruce. 2001. "Tremblay, Gilles." P. 716 in *The New Grove Dictionary of Music and Musicians*, vol. 25, ed. Stanley Sadie. 2nd ed. London: Macmillan.

Maxwell Davies, Peter. 2001. *Symphony No. 8*. Available at www.maxopus.com/works/symph8.htm (accessed May 2001).

Mayer, Harold M., and Richard C. Wade. 1969. *Chicago: Growth of a Metropolis*. Chicago: University of Chicago Press.

McDannell, Colleen, and Bernhard Lang. 1988. *Heaven: A History*. New Haven, CT: Yale University Press.

McLeay, Colin R. 1994. "'The 'Dunedin Sound': New Zealand Rock and Cultural Geography." *Perfect Beat*, no. 2, pp. 38–50.

———. 1995. "Musical Words, Musical Worlds: Geographic Imagery in the Music of U2." *New Zealand Geographer*, vol. 51, no. 2, pp. 1–6.

Meinicke, Bruno. 1948. "Music and Medicine in Classical Antiquity." Pp. 47–95 in *Music and Medicine*, ed. Dorothy M. Schullian and Max Schoen. New York: Schuman.

Meinig, D. W., ed. 1979. *The Interpretation of Ordinary Landscapes*. New York: Oxford University Press.

Melvin, Sheila, and Jindong Cai. 2001. "The Offspring of a Grim Revolution." *New York Times*, April, 1, Section 2, pp. 1, 30.

Meyer, D. K. 1976. "Country Music and Geographical Themes." *Mississippi Geographer*, vol. 4, pp. 65–74.

Meyer, L. B. 1956. *Emotion and Meaning in Music*. Chicago: University of Chicago Press.

———. 1989. *Style and Music: Theory, History, and Ideology*. Philadelphia: University of Pennsylvania Press.

Mikesell, Marvin W. 1969. "The Borderlands of Geography as a Social Science." Pp. 227–48 in *Interdisciplinary Relationships in the Social Sciences*, ed. Muzafer Sherif and Carolyn W. Sherif. Chicago: Aldine.

Millington, Barry, and Stewart Spencer, eds. 1992. *Wagner in Performance*. New Haven, CT: Yale University Press.

Mitcalfe, Barry. 1974. *Maori Poetry: The Singing Word*. Wellington, New Zealand: Victoria University Press.

Montagu, Jeremy. 2002. *Timpani and Percussion*. New Haven, CT: Yale University Press.

Murphy, Alexander B., and Douglas L. Johnson, eds. 2000. *Cultural Encounters with the Environment*. Lanham, MD: Rowman & Littlefield.

Narazaki, Yoko, with Masakata Kanazawa. 2001. "Takemitsu, Tōru." Pp. 22–25 in *The New Grove Dictionary of Music and Musicians*, vol. 25, ed. Stanley Sadie. 2nd ed. London: Macmillan.

Nash, Peter H. 1968. "Music Regions and Regional Music." *The Deccan Geographer*, vol. 6, pp. 1–24.
———. 1974. *Music and Cultural Geography*. Waterloo, ON: University of Waterloo, Faculty of Environmental Studies.
———. 1975a. "Music and Cultural Geography." *The Geographer*, vol. 22, pp. 1–14.
———. 1975b. "Music and Environment: An Investigation of Some of the Spatial Aspects of Production, Diffusion, and Consumption of Music." *Canadian Association of University Schools of Music Journal*, vol. 5, pp. 42–71.
Nash, Peter H., and George O. Carney. 1996. "The Seven Themes of Music Geography." *Canadian Geographer*, vol. 40, no. 1, pp. 69–74.
"National Anthems." 2001. Pp. 655–87 in *The New Grove Dictionary of Music and Musicians*, vol. 17, ed. Stanley Sadie. 2nd ed. New York: Macmillan.
Nettl, Bruno, and Ruth M. Stone, eds. 1998–2002. *The Garland Encyclopedia of World Music*. 10 vols. New York: Routledge.
Northey, Margot, and David B. Knight. 2005. *Making Sense: Geography and Environmental Sciences*. Updated 2nd ed. Toronto, ON: Oxford University Press.
Norton, William. 2000. *Cultural Geography: Themes, Concepts, Analysis*. Toronto, ON: Oxford University Press.
O'Connell, Charles. 1935. *The Victor Book of the Symphony*. New York: Simon and Schuster.
———. 1941. *The Victor Book of the Symphony*. Rev. ed. New York: Simon and Schuster.
———. 1950. *The Victor Book of Overtures, Tone Poems and Other Orchestral Works*. New York: Simon and Schuster.
Orrey, Leslie. 1975. *Programme Music: A Brief Survey from the Sixteenth Century to the Present Day*. London: Davis-Poynter.
Osborne, Richard. 1995. "Beethoven." Pp. 80–106 in *A Guide to the Symphony*, ed. Robert Layton. Oxford, UK: Oxford University Press.
Ottaway, Hugh. 1972. *Vaughan Williams Symphonies*. London: BBC Books.
Page, Tim. 1999. "Karlheinz Stockhausen." Pp. 70–71 in *Perspectives: Maurizio Pollini*, ed. Theodore W. Libbey Jr. New York: Carnegie Hall.
Paine, Anthony. 2001. "Bridge, Frank." Pp. 347–49 in *The New Grove Dictionary of Music and Musicians*, vol. 4, ed. Stanley Sadie. 2nd ed. London: Macmillan.
Partington, John T. 1995. *Making Music*. Ottawa, ON: Carleton University Press.
Paterson, J. L. 1991. "Putting Pop into Its Place: Using Popular Music in the Teaching of Geography." *New Zealand Journal of Geography*, no. 92, pp. 18–19.
Peyser, Joan, ed. 1986. *The Orchestra: Origins and Transformations*. New York: Charles Scribner.
Pocock, Douglas. 1987. "Sounds of a Cathedral City." Durham, UK: University of Durham, Department of Geography, tape.
———. 1988. "The Music of Geography." Pp. 62–71 in *Humanistic Approaches in Geography*, ed. Douglas C. D. Pocock. Department of Geography Occasional Paper no. 22. Durham, UK: University of Durham.
———. 1989. "Sound and the Geographer." *Geography*, vol. 79, no. 13, pp. 193–200.
Ponting, Herbert G. 1921. *The Great White South; or, With Scott in the Antarctic*. London: Duckworth.
Proctor, George. 1980. *Canadian Music of the Twentieth Century*. Toronto, ON: University of Toronto Press.

Ragnarsson, Hjálmar H. 1996. *Jón Leifs: Geyser and Other Orchestral Works*. BIS, CD-830.
Revill, George. 1991. "The Lark Descending: Monument to a Radical Pastoralism." *Landscape Research*, vol. 16, pp. 25–30.
——. 1995. "Hiawatha and Pan-Africanism: Samuel Coleridge-Taylor (1875–1912), a Black Composer in Suburban London." *Ecumene*, vol. 2, no. 3, pp. 247–66.
——. 1998. "Samuel Coleridge-Taylor's Geography of Disappointment: Hybridity, Identity, and Networks of Musical Meaning." Pp. 197–221 in *The Place of Music*, ed. Andrew Leyshon, David Matless, and George Revill. New York: Guilford Press.
——. 2000. "English Pastoral: Music, Landscape, History and Politics." Pp. 140–58 in *Cultural Turns: Geographical Turns*, ed. Ian Cook et al. Harlow, Essex: Prentice Hall.
——. 2004. "Performing French Folk Music." *Cultural Geographies*, vol. 1, pp. 199–209.
——. In press. "Vernacular Culture and the Place of Folk Music." *Social and Cultural Geography*.
Rickards, Guy. 1997. *Jean Sibelius*. London: Phaidon.
Rockwell, John. 1997. *All American Music: Composition in the Late Twentieth Century*. New York: Da Capo.
Russell, Jeffrey B. 1997. *A History of Heaven: The Singing Silence*. Princeton, NJ: Princeton University Press.
Rycroft, Simon. 1998. "Global Undergrounds: The Cultural Politics of Sound and Light in Los Angeles, 1965–1975." Pp. 222–48 in *The Place of Music*, ed. Andrew Leyshon, David Matless, and George Revill. New York: Guilford Press.
Sack, Robert David. 1980. *Conceptions of Space in Social Thought*. Minneapolis: University of Minnesota Press.
Sadie, Stanley, ed. 1990. *Brief Guide to Music*. Englewood Cliffs, NJ: Prentice Hall.
——. 2001. *The New Grove Dictionary of Music and Musicians*, 29 vols. London: Macmillan.
Salmenhaara, Erkki. 1994. *Klami*. Colchester, Essex: Chandos 9268.
Salzman, Eric. 1974. *Twentieth-Century Music*, 2nd ed. Englewood Cliffs, NJ: Prentice Hall.
Schafer, R. Murray. 1969. *The New Soundscape*. Don Mills, ON: BMI Canada.
——. 1977. *The Tuning of the World*. Toronto, ON: McClelland and Stewart.
——. 1984. *On Canadian Music*. Bancroft, ON: Arcana Editions.
——. 1986. *The Thinking Ear*. Toronto, ON: Arcana Editions.
——. 2002. *Patria: The Complete Cycle*. Toronto, ON: Coach House Books.
Schauffler, Robert A. 1947. *Beethoven: The Man Who Freed Music*. New York: Tudor.
Schonberg, Harold C. 1981. *The Lives of the Great Composers*. New York: Norton.
Schwartz, Elliott, and Barney Childs, eds. 1998. *Contemporary Composers on Contemporary Music*. Expanded ed. New York: Da Capo Press.
Schwartz, Elliott, and Daniel Godfrey. 1993. *Music Since 1945*. New York: Schirmer.
Shafer, Boyd C. 1955. *Nationalism: Myth and Reality*. New York: Harcourt.
——. 1972. *Faces of Nationalism*. New York: Harcourt Brace Jovanovich.
Shieff, Sarah. 2002. *Talking Music: Conversations with New Zealand Musicians*. Auckland, New Zealand: Auckland University Press.

Shulman, Laurie. 1997. *Shostakovich Symphony No. 8*. Hollywood: Delos International, DE3204.
Shute, Jeremy J., and David B. Knight. 1995. "Obtaining an Understanding of Environmental Knowledge." *Canadian Geographer*, vol. 39, no. 2, pp. 101–11.
Simpson, Robert. 1991. *The Essence of Bruckner*. London: Victor Gollancz.
Sisman, Elaine. 2000. "Memory and Invention at the Threshold of Beethoven's Late Style." Pp. 51–87 in *Beethoven and His World*, ed. Scott Burnham and Michael P. Steinberg. Princeton, NJ: Princeton University Press.
Slonimsky, Nicolas. 1965. *Lexicon of Musical Invective: Critical Assaults on Composers Since Beethoven's Time*. 2nd ed. New York: Coleman-Ross.
Smith, Susan J. 1994. "Soundscape." *Area*, vol. 26, no. 3, pp. 232–40.
Solti, Georg. 1997. *Memoirs*. New York: A Capella Books.
Spratt, Geoffrey K. 1987. *The Music of Arthur Honegger*. Cork, Ireland: Cork University Press.
Staines, J., et al. 2001. *Classical Music: The Rough Guide*. 3rd ed. London: Rough Guides.
Stannard, Neil. 1999. *Hovhaness Collection*, vol. 2. New York: Delos International, DE3711.
Steinberg, Michael. 1995. *The Symphony*. New York: Oxford University Press.
———. 1998. *The Concerto*. New York: Oxford University Press.
Sternberg, Rolf. 1998. "Fantasy, Geography, Wagner, and Opera." *Geographical Review*, vol. 88, no. 3, pp. 327–48.
Stockhausen, Karlheinz. 1964. *Texte*, vol. 2. Köln, Germany: Verlag M. DuMont Schauberg.
Stradling, Robert. 1998. "England's Glory: Sensibilities of Place in English Music, 1900–1950." Pp. 176–96 in *The Place of Music*, ed. Andrew Leyshon, David Matless, and George Revill. New York: Guilford Press.
Stump, Roger W. 1998. "Place and Innovation in Popular Music: The Bebop Revolution in Jazz." *Journal of Cultural Geography*, vol. 18, pp. 11–34.
Sweeney-Turner, Steve. 1998. "Borderlines: Bilingual Terrain in Scottish Song." Pp. 151–75 in *The Place of Music*, ed. Andrew Leyshon, David Matless, and George Revill. New York: Guilford Press.
Takemitsu, Tōru. 1991. *Tōru Takemitsu, Riverrun . . .* London: Virgin Classics, VC7-91180-2.
———. 1995. *Confronting Silence: Selected Writings*. Berkeley, CA: Fallen Leaf Press.
Taruskin, Richard. 2001. "Nationalism." Pp. 689–706 in *The New Grove Dictionary of Music and Musicians*, vol. 17, ed. Stanley Sadie. 2nd ed. New York: Macmillan.
Tavenor, Jean. 1970. "Musical Themes in Latin American Culture: A Geographical Appraisal." *The Bloomsbury Geographer*, vol. 3, pp. 60–66.
Thayer, Alexander W. 1967. *Life of Beethoven*, ed. Elliot Forbes. Princeton, NJ: Princeton University Press.
Thirwall, Stephen L. 1992. *Musical Landscape: A Definition and a Case Study of Musical Landscape in Its Contribution to the Development of Québecois Identity*. Master's thesis, Carleton University, Ottawa, Canada.
Todd, R. Larry. 1993. *Mendelssohn: The Hebrides and Other Overtures*. Cambridge: Cambridge University Press.

Tovey, Donald Francis. 1935–1939. *Essays in Musical Analysis*, 6 vols. London: Oxford University Press.
———. 1941. *A Musician Talks*, 2 vols. London: Oxford University Press.
Trapp, Maria Augusta. 1990. *The Story of the Trapp Family Singers*. New York: Doubleday.
Tuan, Yi-Fu. 1974. *Topophilia*. Englewood Cliffs, NJ: Prentice Hall.
———. 1977. *Space and Place*. Minneapolis: University of Minnesota Press.
———. 1976. "Humanistic Geography." *Annals of the Association of American Geographers*, vol. 66, pp. 266–76.
———. 1978. "Sign and Metaphor." *Annals of the Association of American Geographers*, vol. 68, pp. 363–72.
———. 1998. *Escapism*. Baltimore: Johns Hopkins University Press.
———. 1999. *Who Am I? An Autobiography of Emotion, Mind, and Spirit*. Madison: University of Wisconsin Press.
Varèse, Edgard. 1998. "Spatial Music." Pp. 204–7 in *Contemporary Composers on Contemporary Music*, ed. Elliot Schwartz and Barney Childs. Exp. ed. New York: Da Capo Press.
Vaughan Williams, Ralph. 1934. *National Music*, reprinted in 1996 in his *National Music and Other Essays*. 2nd ed. New York: Clarendon Press.
———. 1965. *The Making of Music*. Ithaca, NY: Cornell University Press.
Vigeland, Carl A. 1991. *In Concert: Onstage and Offstage with the Boston Symphony Orchestra*. Amherst: University of Massachusetts Press.
Violita, Sister. 1980. "The Geography of Music." Pp. 353–66 in *Recent Trends and Concepts in Geography*, ed. Ram Bahadur Mandal and Vishwa Nath Prasad. Sinha. New Delhi: Concept Publishing.
Wagner, Phillip, and Marvin Mikesell. 1962. *Readings in Cultural Geography*. Chicago: University of Chicago Press.
Wall, Melanie. 2000. "The Popular and Geography: Music and Racialized Identities in New Zealand." Pp. 75–87 in *Cultural Turns: Geographical Turns*. Harlow, Essex: Prentice Hall.
Waterman, Stanley. 1998a. "Carnivals for Élites? The Cultural Politics of Arts Festivals." *Progress in Human Geography*, vol. 22, pp. 54–74.
———. 1998b. "Place, Culture and Identity: Summer Music in Upper Galilee." *Transactions of the Institute of British Geographers*, vol. 23, pp. 253–67.
Waterman, Stanley, and Stanley D. Brunn, eds. In press. *Geography and Music*. Dordrecht, Netherlands: Kluwer-Springer.
Watson, Derek, comp. 1994. *The Wordsworth Dictionary of Musical Quotations*. London: Wordsworth Editions.
Watson, James Wreford. 1979. *Countryside Canada*. Frederick, NB: Fiddlehead Poetry Books.
Whiteley, Sheila, Andy Bennett, and Stan Hawkins. 2004. *Music, Space and Place: Popular Music and Cultural Identity*. Burlington, VT: Ashgate.
Whone, Herbert. 1990. *Church, Monastery, Cathedral: An Illustrated Guide to Christian Symbolism*. Longmead, Dorset: Element Books.
Willoughby, David. 1996. *The World of Music*. 3rd ed. Dubuque, IA: Brown and Benchmark.

Woods, Louis, and Charles F. Gritzner. 1982. "A Conceptual Model of the Pastoral Elements in Country-Western Music." Pp. 172–73 in *AAG Program Abstracts*. Washington, DC: Association of American Geographers.

Woolf, Penelope. 1988. "Symbol of the Second Empire: Cultural Politics and the Paris Opera House." Pp. 214–35 in *The Iconography of Landscape*, ed. Denis Cosgrove and Stephen Daniels. Cambridge: Cambridge University Press.

Wright, John K. 1947. "*Terrae incognitae*: The Place of Imagination in Geography." Pp. 1–15 in *Annals of the Association of American Geographers*, vol. 37. Reprinted in 1966 on pp. 68–88 in John K. Wright, *Human Nature in Geography*. Cambridge, MA: Harvard University Press.

Xenakis, Iannis. 1971. *Formalized Music: Thought and Mathematics in Composition*. Bloomington: Indiana University Press.

Zelinsky, Wilbur. 1976. "Unearthly Delights: Cemetery Names and the Map of the Changing American Afterworld." Pp. 171–96 in *Geographies of the Mind*, ed. David Lowenthal and Martyn J. Bowden. New York: Oxford University Press.

Index

About the Author

David B. Knight is a cultural geographer and orchestral musician who has lived, worked, and performed in New Zealand, Scotland, the United States, and Canada. He has been an instructor in music, a professor of geography, dean of social sciences at the University of Guelph, and director and general editor of the Carleton University Press. He also served for five years as chair/président of the I.G.U. Commission on the World Political Map. He has authored numerous articles and book chapters, including "Geographies of the Orchestra," and has authored, edited, and coedited many books, including *Restructuring Societies* (1999) and *Making Sense: Geography and Environmental Sciences* (2005). Now retired from university life, he remains active as an orchestral musician. He and his wife live in the village of Elora in Ontario, Canada, and enjoy visits to their cottage in Vermont.